Reference Materials in Measurement and Technology

Sergey V. Medvedevskikh · Egor P. Sobina ·
Olga N. Kremleva · Mikhail V. Okrepilov
Editors

Reference Materials in Measurement and Technology

Proceedings of the Fourth International Scientific Conference

Springer

Editors
Sergey V. Medvedevskikh
D. I. Mendeleyev Institute for Metrology (VNIIM)
Saint Petersburg, Russia

Olga N. Kremleva
UNIIM—Affiliated Branch of the D. I. Mendeleyev Institute for Metrology
Yekaterinburg, Russia

Egor P. Sobina
UNIIM—Affiliated Branch of the D. I. Mendeleyev Institute for Metrology
Yekaterinburg, Russia

Mikhail V. Okrepilov
D. I. Mendeleyev Institute for Metrology (VNIIM)
Saint Petersburg, Russia

ISBN 978-3-031-06284-1 ISBN 978-3-031-06285-8 (eBook)
https://doi.org/10.1007/978-3-031-06285-8

Translation from the Russian language edition: "Measurement Standards. Reference Materials." by Sergey V. Medvedevskikh et al., © Author 2021. Published by UNIIM. All Rights Reserved.

This Springer imprint is published by the registered company Springer Nature Switzerland AG
The registered company address is: Gewerbestrasse 11, 6330 Cham, Switzerland

Preface

This Springer book devoted to research performed in the area of reference materials contains selected papers which were presented in December 2020 at the International Scientific Conference 'Reference Materials in Measurement and Technology' held in Saint Petersburg, Russia.

The Conference was organized by the Federal Technical Regulation and Metrology Agency (Rosstandart) in cooperation with the Ural Research Institute of Metrology (UNIIM), an affiliate of the D. I. Mendeleyev All-Russian Research Institute for Metrology (VNIIM). The event provided an open platform for exchanging information about fundamental and applied research regarding the development, production, and application of reference materials. The Conference also aimed to popularize reference materials as the regulatory, procedural, and technical basis for providing traceability and accuracy of measurements.

Measurement accuracy plays a crucial role in all branches of industry and life-supporting spheres. Quality and product safety control is of prime importance in medicine and in clinical and diagnostic studies, in innovative high-tech sectors. At present, the institutes that develop and produce reference materials have to cope with this ever-increasing and becoming more complex demand; nevertheless, the metrological community has learnt to respond to it efficiently.

The task of developing and bringing to the market competitive technological solutions in the shortest possible time, all the while supporting their implementation and application with standards, methodological and regulatory framework turned out to be complicated by the tough period of the COVID-19 pandemic.

We are glad that our Conference managed to bring together successful developers of reference materials of the composition and properties of substances and materials on the one hand and leading experts in various areas of measurement on the other. Professionals from research and metrological institutes, regional metrological centers, enterprises, industrial sector organizations, scientists, faculty members and postgraduates of universities took part in the Conference. The participants represented all Federal districts of Russia. We were also delighted at foreign specialists' involvement: representatives of national metrological services and institutes, ISO/TC 334 (previously ISO/REMCO) and CITAC members, fellow metrologists from Switzerland, Great Britain, Israel, India, Brazil, Ukraine, and Kazakhstan made a valuable contribution to the Conference.

Four special interest groups held their sessions: 'The statutory regulation in the field of reference materials. General issues', 'Reference materials and medical science', 'Quality of life', 'Development of new reference materials, including physical properties and their application'. Fifty-three scientific reports were delivered during these as well as plenary and poster sessions. All of them—directly or indirectly—reflected the interdisciplinary nature of the theory and practice of creating, producing, distributing, and applying reference materials.

The book contains 16 full-size papers delivered at the Conference. All materials underwent the procedure of strict external peer review. The papers presented in this book were published in the Russian language in the peer-reviewed scientific and technical journal 'Measurement Standards. Reference Materials'.

Saint Petersburg, Russia — Sergey V. Medvedevskikh
Yekaterinburg, Russia — Egor P. Sobina
Yekaterinburg, Russia — Olga N. Kremleva
Saint Petersburg, Russia — Mikhail V. Okrepilov

Acknowledgements

On behalf of the Organizing Committee of the 4th International Scientific Conference 'Reference Materials in Measurement and Technology', we would like to thank all invited speakers, session chairpersons, and participants for the successful conduct of our scientific forum. We also wish to express our gratitude to the team of Natalia Popova Laboratory of Scientific Translation for the help with translation and the invaluable advice given in the preparation of this book.

The Conference Organizing Committee would like to extend the sincerest appreciation to V. B. Baranovskaya, I. Ye. Vasilieva, M. S. Vonsky, A. V. Gordeyev, Ye. V. Davydova, A. P. Druchinin, I. N. Zyryanova, A. I. Krylov, M. Yu. Medvedevskikh, Ye. V. Osintseva, A. S. Sergeyeva, V. A. Syasko for their immense help with the reviewing and editing of the papers included in this volume.

The Organizing Committee would like to express special gratitude to N. S. Tarayeva whose help has made this publication possible.

Contents

Calibration of AMV-1006M NMR Analyzers Using Natural Seeds and Imitator Reference Materials

Oleg S. Agafonov and Sergey M. Prudnikov

Abstract The present article studies the metrological characteristics of AMV-1006M NMR analyzers obtained during their calibration to determine the mass fraction of oleic acid in sunflower oil. Calibration is of great importance for obtaining repeatable and reproducible measurements of the quality indicators of agricultural raw materials using measuring instruments. Calibration can be carried out in different ways. In this case, the calibration procedure is performed using sunflower seeds and specially developed imitator reference materials (RMs) for the mass fraction of oleic acid in seed oil. The developed imitator RMs are characterized by high long-term stability of their physicochemical parameters and do not require special storage and operating conditions. The obtained data confirm the feasibility of using these imitator RMs for calibration purposes. In the calibration of analyzers, the random error limits of measurement results when using natural seeds exceeded those obtained with the use of imitator RMs by a factor of 3. Furthermore, the study of the temperature effect on calibration results indicates that the standard deviation of measurement results does not exceed 0.8 abs% for the imitator RMs, whereas the use of natural seeds yields 2.5 abs%.

Keywords Sunflower seeds · Oleic acid mass fraction · Nuclear magnetic relaxation · Imitator reference materials · Calibration · Metrological support

Introduction

Stable development of present-day companies requires the introduction of control systems implementing feedback principles. Such systems improve the quality of end products and guarantee their consistent quality. According to regulatory documents [1], companies engaged in the fat-and-oil industry are to monitor oil and moisture content during sunflower seed harvesting, which directly affects oil yield, as well as storage and processing conditions. Furthermore, the emergence of new raw materials,

O. S. Agafonov (✉) · S. M. Prudnikov
V.S. Pustovoit All-Russian Research Institute of Oil Crops, Krasnodar 350038, Russia
e-mail: sacred_jktu@bk.ru

S. V. Medvedevskikh et al. (eds.), *Reference Materials in Measurement and Technology*,
https://doi.org/10.1007/978-3-031-06285-8_1

specifically high-oleic sunflower seeds [2], has set a new challenge for companies to control oleic acid content in seed oil. Currently, this indicator is not regulated by state standards, so it is set by companies instead; the minimum baseline value varies from 78 to 85%, depending on the company's requirements.

The mass fraction of oleic acid in sunflower oil is significantly influenced by species-specific characteristics, as well as by growing conditions for seeds [3–5]. In addition, deliberate seed adulteration is possible due to the higher cost of such raw materials. The above-mentioned factors necessitate companies to exercise prompt control over this indicator.

Presently, the chromatographic method is used as an umpire method to determine the mass fraction of oleic acid in sunflower oil [6, 7]. The disadvantages of this method include its poor performance, high labor intensity, high qualification requirements for staff, as well as the need for consumables. The listed disadvantages do not allow the chromatographic method to be used for prompt control over the mass fraction of oleic acid in sunflower oil.

Literary sources describe rapid methods using refractometry [8] and IR-spectroscopy[1] [9–11] that are difficult to implement at factory laboratories. In addition, these methods exhibit low representativeness, require a lot of time to implement and obtain calibration curves, as well as improve them in the process of operation, with the correlation coefficients not exceeding 85%.

Earlier, we developed an instrumental method for determining the mass fraction of oleic acid in sunflower oil [12] using the pulsed NMR[2] method. The advantages of this method consist in its rapidness (analysis time of one sample – below 1 min), simple sample preparation, high representativeness, and possible simultaneous determination of several indicators.

Procedures for calibrating analyzers are of great importance in ensuring high repeatability and reproducibility of measurement results via indirect instrumental methods. However, no universal calibration procedures are currently available for measuring instruments, with existing regulatory documents merely serving as guidelines. Moreover, measuring instruments comprise a considerable nomenclature.

An analysis of literary sources has shown that the following basic methods are used for calibrating quantitative analyzers in the fat-and-oil industry:

- use of natural oilseeds [13, 14]. Since they comprise complex biological entities and their physicochemical parameters change during storage, an obvious disadvantage lies in their low long-term stability;
- a type of calibration using natural materials that involves producing reference materials (RMs) from such natural components as grist and oil [15]. The advantage of such RMs over natural seeds consists in the possibility of obtaining repeatable characteristics, whereas their disadvantage lies in the fact that they require special storage conditions and are difficult to produce;

---

[1] IR-spectroscopy – Infrared spectroscopy.

[2] NMR – nuclear magnetic resonance.

Table 1 Fatty acid composition of oil in sunflower seeds used for calibration

Sample	Mass fraction of fatty acids in sunflower oil, % of total fatty acids					
	Palmitic acid C 16:0	Stearic acid C 18:0	Oleic acid C 18:1	Linoleic acid C 18:2	Linolenic acid C 18:3	Other
1	5	3.6	31.1	60.1	0.1	0.1
2	5.9	3.9	47.6	41.2	0.2	1.2
3	5.2	2.1	68.9	18.5	3.9	1.4
4	5.7	6.7	72.1	13.9	0.1	1.5
5	4.3	3.7	79	11.2	0.4	1.4
6	4.3	3.7	86.2	5.5	0.1	0.2

- use of imitator RMs produced from chemically inert substances. They provide metrological characteristics within a given range, require no special storage conditions, and exhibit high long-term stability [16, 17].

For calibrating AMV-1006M NMR analyzers,[3] we developed a special imitator RM set for the mass fraction of oleic acid in sunflower oil [18].

The present article aims to study the influence of used materials on the calibration results of NMR analyzers when determining the mass fraction of oleic acid in sunflower oil. The obtained results will provide a practical justification for the effectiveness of imitator RMs in calibrating quantitative NMR analyzers.

Materials and Methods

The studies were carried out at the Central Experimental Base (CEB) of VNIIMK (Krasnodar) in 2017–2020. The sample preparation of sunflower seeds was conducted in accordance with GOST 8.597–2010 [19] used to determine the oil and moisture content of oil crops employing the pulsed NMR method [20]. The sample preparation procedure consisted in collecting a portion of sunflower seeds weighing (150 ± 5) g from the material under analysis, as well as 3-mm sieving the sample to remove foreign matter. Prior to performing measurements, each sample was checked for the presence of metal impurities. The prepared samples were stored in tightly closed containers at 23 °C. Oleic acid measurements were performed using a quantitative NMR analyzer AMV-1006M (VNIIMK, Krasnodar). This analyzer is used at most Russian and CIS fat-and-oil companies for promptly determining the oil and moisture content of oil crops and their products. The analyzer was upgraded to determine oleic acid content using special software [20].

The oil of sunflower seeds prepared for the study exhibited a mass fraction of oleic acid ranging from 31 to 86% (Table 1).

[3] AMV-1006M NMR analyzers, Federal Information Fund for Ensuring the Uniformity of Measurements. Available at: https://fgis.gost.ru/fundmetrology/registry/4/items/325229.

Table 2 Characteristics of imitator RMs for the mass fraction of oleic acid in sunflower oil

RM number	Certified value–oleic acid mass fraction imitated in the RMs, %	Absolute error limits of certified RM values ($P = 0.95$), %
1	31	±1
2	52	
3	62	
4	73	
5	81	
6	93	

Oil was extracted from the prepared seed samples using a laboratory press. Then the resulting oil was filtered to remove impurities and dissolved in hexane, followed by the addition of a fresh solution of sodium ethoxide in methanol, having a concentration of 2 mol/dm^3. After intensive stirring followed by settling, the resulting mixture was paper filtered [6].

The fatty acid composition of sunflower oil was determined using a Chromatec-Crystal 5000 chromatograph (JSC SDO Chromatec, Yoshkar-Ola). The instruction manual of the chromatograph was used to prepare the analyzer for operation and analyze the prepared methyl esters.

Imitator RMs for the mass fraction of oleic acid in sunflower oil were developed and patented at VNIIMK to simplify the process of calibrating quantitative AMV-1006M NMR analyzers [21]. These instruments comprise hermetically sealed ceramic ampoules filled with a mixture of organosilicon liquids that imitate different oleic acid contents. The weighted average of the proton spin–spin relaxation time of the used organosilicon liquids ranges from 100 to 200 ms. Table 2 presents the mass fractions of oleic acid in sunflower oil for imitator RMs.

The mass fractions of oleic acid in sunflower oil were certified for the developed imitator RMs using the upgraded AMV-1006M NMR analyzer. This unit is only hypothetically standard as the present procedure is currently being developed and has not been approved yet. The above analyzer was calibrated using natural sunflower seeds, whose mass fractions of oleic acid were determined in accordance with GOST 31,663–2012 *Vegetable Oils and Animal Fats. Mass Fraction Determination of Fatty Acid Methyl Esters* via *Gas Chromatography* [7]. According to the specified GOST, the repeatability of measurement results did not exceed 1 abs%. In order to reduce the random component of measurement error, the average mass fraction of oleic acid obtained in ten parallel measurements was used for each seed sample.

In most cases, fat-and-oil companies apply GOST 30,418–96 reducing the accuracy of results due to a greater error of up to 5 abs% [6].

The imitator RMs were used to calibrate the other NMR analyzers specified in the study. This study was conducted to show the advantages of these RMs for the calibration of other (hypothetically working) AMV-1006M NMR analyzers in order to ensure the reproducibility of calibration results, as well as the uniformity of further measurement results, for all working NMR-analyzers. Future certification of

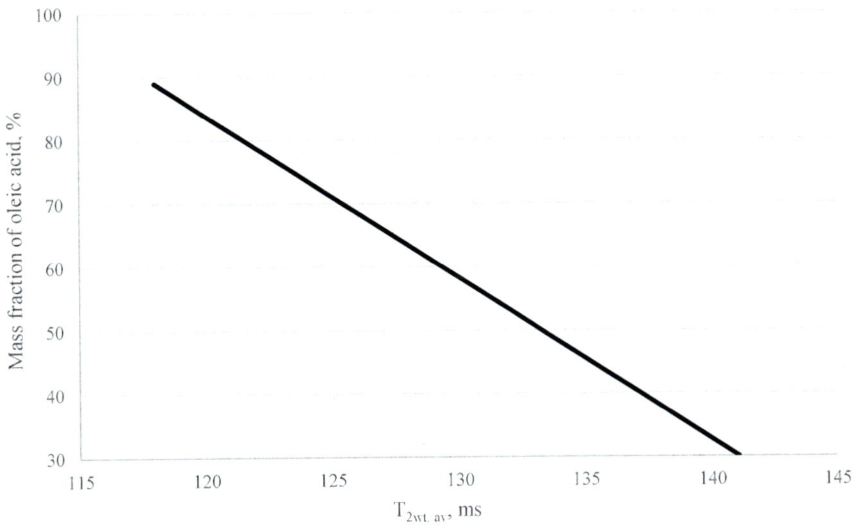

Fig. 1 Calibration curve for determining the mass fraction of oleic acid in sunflower oil using the pulsed NMR method

all imitator RM sets is expected to be conducted using a single unit (hypothetically standard unit).

When measuring nuclear magnetic relaxation characteristics, the material under analysis was thermostated at the required temperature for 2 h.

The calibration curve obtained using natural sunflower seed is shown in Fig. 1. The presented data indicate a linear relationship between the nuclear magnetic relaxation characteristics of oleic acid protons and oleic acid content in seed oil, with the correlation coefficient exceeding 0.991. Furthermore, this linear character is observed over the entire considered range of oleic acid content in sunflower oil.

The data obtained during the study were processed using the methods of mathematical and physical modeling, statistical processing, as well as interpolation and correlation analysis from the Mathcad. 8 (Professional) and Mathlab 5.1 Relaxsometr software package.

Results and Discussion

The first stage of the study produced the calibration repeatability results of one AMV-1006M NMR analyzer using imitator RMs and natural seeds. Table 3 shows the measurement results of imitator RMs and sunflower seeds having different certified values of oleic acid mass fraction; these measurements were replicated 12 times using the same NMR analyzer.

Table 3 Repeatability of metrological characteristics obtained during the calibration of the AMB-1006M NMR analyzer using imitator RMs and natural seeds

Measurement	Measured oleic acid mass fraction, %			
	Imitator RMs		Seeds	
	31	**81**	**31**	**79**
1	30	80	30	77
2	30	80	29	75
3	31	80	31	78
4	30	82	34	79
5	30	80	28	77
6	32	81	32	76
7	31	82	32	79
8	30	80	30	76
9	31	82	35	82
10	32	81	36	81
11	31	81	35	78
12	32	80	34	83
$\overline{x}$	**30.8**	**80.8**	**32.2**	**78.4**
$S_{\overline{x}}$	**0.25**	**0.26**	**0.79**	**0.75**
ε	**0.57**	**0.59**	**1.79**	**1.71**

The relative standard deviations of the measurement results ($S_{\overline{x}}$), as well as the confidence random error limits of the measurement results (ε), are almost three times lower for the measurement results obtained using the imitator RMs as compared to the use of sunflower seeds.

In order to check the reproducibility of calibration results obtained using the imitator RMs and sunflower seeds, the study employed four AMV-1006M NMR analyzers operating under the same conditions to eliminate the temperature effect on calibration results (Tables 4 and 5).

The prepared seed samples and the imitator RMs for the mass fraction of oleic acid were thermostatted at 23 ± 0.5 °C for 2 h prior to each calibration.

It follows from Tables 4 and 5 that the calibration results of NMR analyzers using the imitator RMs exceed those obtained using natural sunflower seed in terms of reproducibility almost by a factor of three.

Our studies indicate that the imitator RMs provide higher metrological characteristics of NMR-analyzer calibration as compared to the use of natural seeds as reference materials.

This factor is attributable to the homogeneous structure of imitator RMs formed in the process of their production. Sunflower seeds constitute a complex heterogeneous system, exhibiting structural heterogeneity even within a single sample under analysis.

Table 4 Measurement reproducibility when calibrating NMR analyzers with the use of the imitator RMs for the mass fraction of oleic acid

RM	Certified value, %	Mass fraction of oleic acid measured using four analyzers, %				$S_{\overline{x}}$	ε
		1	2	3	4		
1	31	30.4	31.1	31.6	32.3	**0.40**	**1.28**
2	52	51.7	53.2	52.6	51.6	**0.38**	**1.21**
3	62	61.4	62.7	62.5	60.9	**0.43**	**1.38**
4	73	72.3	73.4	73.7	74.0	**0.37**	**1.18**
5	81	81.9	79.9	80.2	80.7	**0.44**	**1.40**
6	93	93.8	94.1	92.6	94.3	**0.38**	**1.21**

Table 5 Measurement reproducibility when calibrating NMR analyzers with the use of sunflower seeds

Sample	Certified value, %	Mass fraction of oleic acid measured using four analyzers, %				$S_{\overline{x}}$	ε
		1	2	3	4		
1	31	28.3	32.4	31.9	33.6	**1.14**	**3.63**
2	48	45.3	45.1	46.7	49.7	**1.06**	**3.38**
3	69	71.7	65.8	71.6	70.8	**1.41**	**4.47**
4	72	76.2	73.5	73.4	70.3	**1.21**	**3.83**
5	79	75.8	82.3	80.4	80.9	**1.41**	**4.48**
6	86	83.3	84.2	83.4	87.8	**1.06**	**3.37**

In addition, a study of the temperature effect on the calibration results of NMR analyzers was carried out. This study was conducted in the operating room of the AMV-1006M NMR analyzer, where temperature varied within the range of 23 ± 1 °C. This regime corresponded to ordinary conditions at factory laboratories using split systems. The analyzer was calibrated every 30 min for 12 h at actual room temperature measured using a digital thermometer (Fig. 2). The color of the marker indicates the type of used material: black – imitator RMs for the mass fraction of oleic acid, gray–natural sunflower seeds.

The presented data indicate that the nuclear magnetic relaxation characteristics of oil protons in seeds are affected by temperature to a greater extent than analogous characteristics exhibited by the protons of the imitator RMs for the mass fraction of oleic acid. The standard deviation of measurement results does not exceed 0.8 abs% for the imitator RMs, whereas the use of natural seeds yields 2.5 abs%. Such nature of the temperature effect is attributable to the fact that temperature has a smaller effect on the mobility of protons contained in imitator substances than in natural seeds.

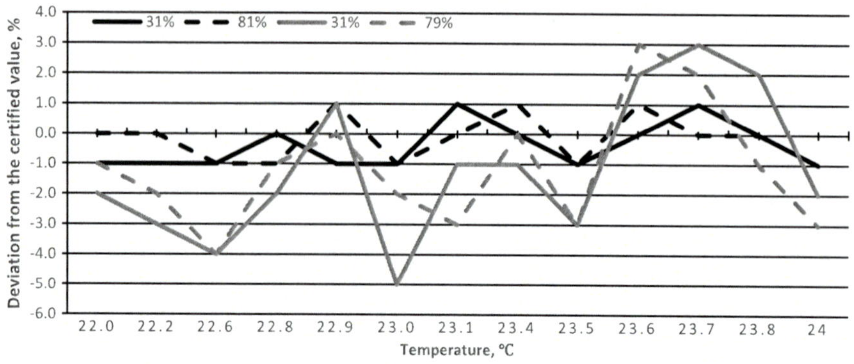

Fig. 2 Calibration error of the AMB-1006M NMR analyzer at different temperatures

Conclusion

The present study involved a metrological evaluation of materials used in the calibration of quantitative NMR analyzers of oleic acid mass fraction in sunflower oil in terms of their influence on calibration repeatability and reproducibility. A comparison was made between the use of natural sunflower seeds and imitator RMs for the mass fraction of oleic acid in sunflower oil developed by the present authors.

The study indicates that the imitator RMs for the mass fraction of oleic acid in sunflower oil can improve the repeatability and reproducibility of calibration results almost threefold. This enhancement is achieved due to the high homogeneity of these imitator RMs. Such homogeneity results from their production process, as well as from the used substances imitating nuclear magnetic relaxation characteristics.

The developed imitator RMs are less sensitive to temperature effects, as well as exhibiting higher stability of their physicochemical parameters and, hence, their certified values. An essential factor consists in the simplicity of using the imitator RMs as compared to natural seeds. The RM set was developed and patented to imitate oleic acid mass fraction in sunflower oil within the range of 31% to 93% [18].

Since the present work discusses only the calibration repeatability and reproducibility of quantitative NMR analyzers using imitator RMs and natural sunflower seeds, the next step is to examine the validity of measured oleic acid mass fractions in sunflower oil using the pulsed NMR method. These studies have been conducted from 2019 to the present day at Russian fat-and-oil companies: Uryupinsk and Sorochinsk Elevators, Bolsheglushitskoe HPP (Samara Region), Olsam LLC, and Pavlovskagroproduct JSC (Voronezh Region).

Author Contributions Both authors have contributed equally to the preparation of this article.

Conflict of Interest The article was prepared on the basis of a report presented at the IV International Scientific Conference "Reference Materials in Measurement and Technology" (St. Petersburg, December 1–3, 2020). The article was admitted for publication after the abstract was revised, the article was formalized and the review procedure was carried out.

The version in the Russian language is published in the journal "Measurement Standards. Reference Materials" 2021;17(1):47–57. https://doi.org/10.20915/2687-0886-2021-17-1-47-57.

References

1. GOST 22391–2015 (2020) Sunflower, Specifications. Standartinform, Moscow, 13p (In Russ)
2. Four types of sunflower oil. https://sunflowernsa.com/oil/Four-Types-of-Sunflower-Oil/
3. Fadul O (2012) Changes in oil fatty acid composition during seed development of sunflower. Asian J Plant Sci 11:241–245. https://doi.org/10.3923/ajps.2012.241.245
4. van der Merwe R, Labuschagne MT, Herselman L, Hugo A (2015) Effect of heat stress on seed yield components and oil composition in highand mid-oleic sunflower hybrids. S Afr J Plant Soil 32(3):121–128. https://doi.org/10.1080/02571862.2015.1018354
5. Akkaya M, Çil A, Yücel H, Kola O (2019) The influence of sowing dates on the oil content and fatty acid composition of standard, mid-oleic and high-oleic types of sunflower (Helianthus annuus L.). Food Sci Technol Int 39:448–453. https://doi.org/10.1590/fst.20118
6. GOST 30418–96 (2013) Vegetable oils. Method for determination of fatty acid content. Minsk, 7p (In Russ)
7. GOST 31663-2012 (2019) Vegetable oils and animal fats. Determination of methyl esters of fatty acids by gaz chromatography method. Standartinform, Moscow (In Russ)
8. GOST 28238–89 (2010) Sunflower. Method for determination of oleic acid mass fraction according to the oil refraction index. Standartinform, Moscow, 6p (In Russ)
9. Cantarel MA, Funes IG, Marchevsky EJ, Camiña JM (2009) Determination of oleic acid in sunflower seeds by infrared spectroscopy and multivariate calibration method. Talanta 80(2):489–492. https://doi.org/10.1016/j.talanta.2009.07.004
10. Efimenko SG, Efimenko SK (2018) Express-evaluation of oleic and linoleic fatty acids content in oil in milled sunflower kernels by means of IR-spectrometry. Oil Crops 176(4):58–63 (In Russ.)
11. Pérez-Vich B, Velasco L, Fernández-Martínez J (1998) Determination of seed oil content and fatty acid composition in sunflower through the analysis of intact seeds, husked seeds, meal and oil by near-infrared reflectance spectroscopy. J Am Oil Chem Soc 75(5):547–555. https://doi.org/10.1007/s11746-998-0064-1
12. Agafonov OS, Prudnikov SM, Zverev LV, Rusnak GV (2015) Application of the NMR method to determine the content of oleic acid in sunflower seed oil. In: Development of biological and post-genomic technologies for assessing the quality of agricultural raw materials and creating healthy food products: materials of the 18th international scientific and practical conference dedicated to the memory of V. M. Gorbachev. Moscow, pp. 24–27 (In Russ)
13. The minispec Oil content and moisture in seeds and nuts. http://spectrante.ru/images/pdf_series/Zernovie.pdf
14. SRM 10899–2017 Reference material of mass fraction of fat (oil content) in sunflower seeds (set). https://fgis.gost.ru/fundmetrology/registry/19/items/389300 (In Russ)
15. Kharchuk O, Kuznetsov E, Kirilov A, Geru I (2013) On the issue of standardization of determining the oil content in soybean seeds by NMR relaxation. In: Botnari V, Chișinău (eds) Biotehnologii avansate–realizări şi perspective: Simpozionul ştiinţific naţional cu participare internaţională. Republica Moldova, Tipografia Academiei de Ştiinţe a Moldovei, 48p (In Russ)
16. Prudnikov SM, Agafonov OS, Zverev LV, Vitiuk BI, Gorelikova IA (2001) Oil with the oleinic acid in sunflower seeds various weight fraction protons nuclear magnetic resonance free precession and spin echo signals simulator. Patent RF no 2001610425 (In Russ)

17. SRM 3107–84 Reference material of oil content and humidity of sunflower seeds (set). https://fgis.gost.ru/fundmetrology/registry/19/items/393823 (In Russ)
18. Agafonov OS, Prudnikov SM (2019) Development of simulator reference materials of oleic acid mass fraction in sunflower seed oil. Ref Mater 15(4):42–48. https://doi.org/10.20915/2077-1177-2019-15-4-41-48 (In Russ)
19. GOST 8.597–2010 (2011) State system for ensuring the uniformity of measurements. Oilseeds and oilseeds residues, determination of oil and moisture content using pulsed nuclear magnetic resonance spectrometry. Standartinform, Moscow (In Russ)
20. Prudnikov SM, Zverev LV, Dzhioev TE (2001) A system for receiving and processing signals from pulsed relaxometers of nuclear magnetic resonance. Certificate of official registration of the computer program no 2001610425. Moscow (In Russ)
21. Prudnikov SM, Agafonov OS, Zverev LV, Vitiuk BY, Gorelikova IA (2019) Simulator of a free procession of nuclear magnetic resonance and spin echo oil with a different mass fraction of oleic acid in sunflower seeds. Patent RF no. 2677644 (In Russ)

Oleg S. Agafonov Cand. Sci. (Eng.), Senior Researcher at the Department of Physical Methods of Research, V.S. Pustovoit All-Russian Research Institute of Oil Crops.

Sergey M. Prudnikov Dr. Sci. (Eng.), Professor, Head of the Department of Physical Methods of Research, V.S. Pustovoit All-Russian Research Institute of Oil Crops.

Development of Reference Materials for Gas Permeability

Ilya P. Aronov and Egor P. Sobina

Abstract The article discusses the development of certified reference materials for the gas permeability of rocks with traceability to GET 210–2019 State Primary Standard of Units of Specific Adsorption of Gases, Specific Surface Area of Pores, Specific Volume of Pores, Dimension of Pores, Open Porosity, and Coefficient of Gas Permeability of Solid Substances and Fabricated Materials. At the beginning of the study, the metrological support system for measuring the gas permeability coefficient comprised only certified (type-approved) reference materials (CRMs) with traceability to the measurement results obtained in an interlaboratory experiment using gas permeability analyzers calibrated by means of the same CRMs; at this time, no CRMs traceable to the State Primary Standard were available. The consequent lack of a time-stable comparison basis in the metrological practice of gas permeability measurement prompted the authors to begin work in this area. Key stages in CRM development are considered: analysis of similar CRMs; selection of starting material; experimental studies and the determination of the metrological characteristics of reference materials on their basis. Alumina ceramic cylinders are used as the CRM starting material; the procedure for measuring the gas permeability coefficient is based on the stationary filtration method. The gas permeability coefficient is measured for specimens having different gas permeability values using nitrogen and helium gases. The results of these measurements are used to calculate the absolute gas permeability coefficient, as well as the individual gas permeability coefficients for each specimen, at given back pore pressure values. As a result of the study, a set of reference materials for rock gas permeability (imitators) was approved (GSO 11,546-2020 / GSO 11,550-2020). The certified values of these reference materials range from $0.1 \cdot 10^{-3}$ to 5 μm^2, with the expanded uncertainty of certified values U_o (at $k = 2$ and $P = 0.95$) set at 3%. The developed CRM set can be used to ensure metrological traceability and the reliability of gas permeability coefficient measurements, finding practical applications at petrophysical research laboratories.

I. P. Aronov (✉) · E. P. Sobina
UNIIM–Affiliated Branch of the D. I. Mendeleyev Institute for Metrology, Yekaterinburg, Russia
e-mail: AronovIP@uniim.ru

E. P. Sobina
e-mail: 251@uniim.ru

S. V. Medvedevskikh et al. (eds.), *Reference Materials in Measurement and Technology*,
https://doi.org/10.1007/978-3-031-06285-8_2

Keywords Oil industry · Reference material · Gas permeability coefficient · Absolute gas permeability coefficient · Stationary filtration method · Darcy's Law · Klinkenberg effect

Introduction

In the extractive industries, both in Russia and internationally, one of the areas of focus consists in the provision of metrological support to ensure reliable measurement results during oil-and-gas extraction. The primary physical measurand of these industries is rock permeability [1]. Permeability refers to the property of rocks to transmit liquids and gases, as well as their mixtures, in the presence of a pressure gradient [2]. Almost all core analysis laboratories study permeability, whose measurement is specified as mandatory. This parameter enables the estimation both of the recoverable reserves of hydrocarbon materials and oil-and-gas production technologies, as well as that of the flow rates and productivity of wells [3]. Here, a distinction is made between absolute, effective, and relative permeability. The general characterization of physical properties exhibited by rocks employs absolute permeability, commonly understood to mean the permeability of a porous medium determined in the presence of a single phase that is chemically inert with respect to the rock. Absolute permeability is typically determined either using air or a gas such as helium or nitrogen. Thus, metrology recognizes the measurand of gas permeability (also called the gas permeability coefficient), i.e., the property of rocks that allows the passage of gas in the presence of a pressure differential due to void connectivity (pores, channels, and cracks). An analysis of the Federal Information Fund for Ensuring the Uniformity of Measurements reveals that all existing domestic certified reference materials (CRMs) of rock gas permeability (GSO 8956-2008,[1] GSO 10,799-2016,[2] GSO 10,913–2017/GSO 10,915-2017[3]) were approved according to the results of interlaboratory experiments [4]. However, instead of direct traceability to the State Primary Standard, traceability is only provided to the interlaboratory experiment results obtained using the gas permeability analyzers of rocks calibrated against the same reference materials (RMs). The consequent lack of a time-stable comparison basis in the metrological practice of gas permeability measurement justifies the relevance of the present work.

[1] GSO 8956-2008 Certified Reference Material of the Open Porosity and Gas Permeability of Rocks, Federal Information Fund for Ensuring the Uniformity of Measurements, available at: https://fgis.gost.ru/fundmetrology/registry/19/items/391229.

[2] GSO 10799-2016 Certified Reference Material of Rock Gas Permeability (imitator) (GP CRM UNIIM), Federal Information Fund for Ensuring the Uniformity of Measurements, available at: https://fgis.gost.ru/fundmetrology/registry/19/items/389385.

[3] GSO 10913-2017/GSO 10915-2017 Certified Reference Materials of Open Porosity and Rock Gas Permeability (OPGP CRM set GEOLOGIKA), Federal Information Fund for Ensuring the Uniformity of Measurements, available at: https://fgis.gost.ru/fundmetrology/registry/19/items/389286.

Gas permeability is measured in accordance with GOST 26,450.2-85 [5]. In this study, we use the stationary filtration method due to its direct implementation of Darcy's law. A similar measurement procedure is also described in a Russian paper [6] and applicable foreign standards [7, 8].

In the present study, the helium and nitrogen gas permeability coefficients of the considered RMs were determined using the equipment comprised by GET 210-2019[4] [9]. The uncertainties of the obtained gas permeability coefficients were calculated in accordance with GOST 34,100.3-2017/ISO/IEC Guide 98-3:2008. *Uncertainty of Measurement. Part 3. Guide to the Expression of Uncertainty in Measurement* [10]. Subsequently, a set of rock gas permeability RMs was approved (GSO 11,546–2020/GSO 11,550–2020[5]). The certified values of these RMs range from $0.1 \cdot 10^{-3}$ to 5 μm^2; the expanded uncertainty of certified values U_o (at $k = 2$ and $P = 0.95$) is set at 3%. Such accuracy is sufficient for the metrological support of existing devices designed to determine the gas permeability coefficient with a relative measurement error of 6–30%.

Materials and Methods

Equipment

The RM characterization was carried out using a standard unit belonging to GET 210–2019, which implements the method of stationary filtration. The unit includes the following equipment:

- MG2P500[6] analyzer of porosity and gas permeability (Sanches Technologies Division of Core Laboratories, France): measurement ambit of gas permeability coefficient from $0.1 \cdot 10^{-3}$ to 5 μm^2; pressure measurement ambit from 0 to 1 MPa, with relative error limits of pressure measurement amounting to $\pm$ 0.01% at $P =$ 0.95; temperature measurement ambit from 263.15 to 400 K, with temperature measurement error limits reaching $\pm$ 0.2 K at $P = 0.95$; measurement range of gas volumetric flow rate from 0.02 to 8 dm^3/min under standard conditions (p_0 $= 0.101325$ MPa and $T_0 = 273.15$ K), with relative error limits of volumetric

[4] GET 210-2019 State Primary Standard of Units of Specific Adsorption of Gases, Specific Surface Area of Pores, Specific Volume of Pores, Dimension of Pores, Open Porosity, and Coefficient of Gas Permeability of Solid Substances and Fabricated Materials, Federal Information Fund for Ensuring the Uniformity of Measurements, available at: https://fgis.gost.ru/fundmetrology/registry/12/items/1365157.

[5] GSO 11546-2020/GSO 11550-2020 Certified Reference Material of Rock Gas Permeability (imitators) (GP CRM set)), Federal Information Fund for Ensuring the Uniformity of Measurements, available at: https://fgis.gost.ru/fundmetrology/registry/19/items/1380970.

[6] MG2P500 Analyzers of Porosity and Rock Gas Permeability (State Register Number 68446-17), Federal Information Fund for Ensuring the Uniformity of Measurements, available at: https://fgis.gost.ru/fundmetrology/registry/4/items/382500.

flow rate measurement amounting to ± (0.5% of the reading + 0.1% of the upper measurement limit) at $P = 0.95$;
- digital micrometer 293-241-30[7] (Mitutoyo Corporation, Japan): length measurement ambit from 25 to 50 mm, with error limits of ± 2 μm at $P = 0.95$;
- SOSpr-2b-2–010 mechanical stopwatch[8] (Zlatoust Watch Factory, Russia): time measurement ambit from 0 up to 3600 s, with error limits of ± 0.6 and ± 1.8 s within the ambits of 0–600 s and 600–3600 s, respectively, at $P = 0.95$;
- Thermocenter TC40 Drying Oven (SalvisLab, Switzerland): temperature maintenance ambit from 50 to 200 °C, with a deviation from the set temperature not exceeding ± 0.5 °C;
- CENTER 313 Humidity Temperature Meter (CENTER Technology Corp., Taiwan, FIF No. 22129 09): temperature measurement ambit from 10 to 35 °C, with measurement error limits amounting to ± 0.7 °C at P = 0.95; relative humidity measurement ambit from 10 to 90%, with measurement error limits reaching ± 2.5% at $P = 0.95$;
- desiccator (Version 2) in accordance with GOST 25,336[9] *Laboratory Glassware and Equipment. Basic Parameters and Dimensions*;
- high-purity gaseous helium (compressed), grade A according to TU 20.11.11-001-37,924,839-2019 *High-Purity Gaseous Helium*[10];
- extra pure gaseous nitrogen according to GOST 9293-74.[11]

CRM Material

The following criteria guided the choice of material for gas permeability CRMs. The material should imitate natural rocks as much as possible, be strong, and have a cylindrical shape. Of the analyzed rock gas permeability CRMs, the material comprising GSO 10,799-2016 was found to be the most suitable. This material is characterized by the following parameters: diameter and height of cylinders—approx. 30 mm; material – molded sintered alumina-based refractory with Al_2O_3 mass fraction of at least 85% (EkogeosProm, Russia). The primary advantage of this material over natural rock consists in its greater strength and consequent greater stability. In addition, the manufacturer of this material possesses the technological ability to alter the permeability of the product by changing the fraction size of sintered alumina powder,

[7] Digital micrometer 293-241-30 (State Register Number 30740–12), Federal Information Fund for Ensuring the Uniformity of Measurements, available at: https://fgis.gost.ru/fundmetrology/registry/4/items/338492.

[8] SOSpr-2b-2-010 30 Mechanical Stopwatches (State Register Number 11519–11), Federal Information Fund for Ensuring the Uniformity of Measurements, available at: https://fgis.gost.ru/fundmetrology/registry/4/items/306186.

[9] GOST 25,336 Laboratory Glassware and Equipment. Basic Parameters and Dimensions (with Amendments No 1–4).

[10] TU 20.11.11-001-37,924,839-2019 High-Purity Gaseous Helium.

[11] GOST 9293-74 Gaseous and Liquid Nitrogen. Specifications, Standartinform, Moscow (2007).

thus allowing the development of reference materials within the entire considered range of $(0.1 \cdot 10^{-3}–5)\ \mu m^2$.

Theoretical Background

When determining the metrological characteristics of RMs, the applied measurement procedure is based on the method specified in GOST 26,450.2 [5], i.e., the stationary filtration method. This method consists in determining a constant (stationary) rate of linear gas filtration through a rock specimen under a specified pressure differential. This rate is determined by the known volume of gas that has passed through the specimen over a fixed period of time at a constant pressure differential. Generally, the filtration process for both gases and liquids is described by Darcy's Law

$$Q = \frac{K \cdot S \cdot (p_2 - p_1)}{\mu \cdot l}, \tag{1}$$

where Q—volumetric flow rate of the fluid, m^3/s; S – cross-sectional area of the specimen, m^2; K—permeability coefficient, m^2; p_2 and p_1—pressures at the specimen inlet and outlet, respectively, Pa; μ—fluid viscosity, Pa·s; l—specimen length, m.

Hence, the expression for the permeability coefficient of a cylindrical specimen takes the form:

$$K = \frac{Q \cdot \mu \cdot l}{\pi \cdot \frac{d^2}{4} \cdot (p_2 - p_1)}, \tag{2}$$

where d—specimen diameter, m.

When applying Eq. (2) to the process of gas filtration, it is necessary to take gas compressibility into account. In other words, the volume of gas passing through the specimen at different pressures will also vary. Therefore, the gas permeability coefficient is calculated using the gas volumetric flow rate at the average (pore) pressure in the specimen $P_{por} = (p_2 + p_1)/2$ (pressure in the middle of the specimen). Thus, an equation formulated on the basis of the ideal gas law establishes the relationship of the gas volumetric flow rate Q_{meas} measured at some pressure p_{meas} and temperature T_{meas} with the volumetric flow rate of the same amount of gas Q_{por} at pressure p_{por} and temperature T_{por} in the middle of the specimen.

$$\frac{Q_{meas} \cdot p_{meas}}{T_{meas}} = \frac{Q_{por} \cdot (p_1 + p_2)}{2 \cdot T_{por}} \tag{3}$$

whence it follows that the gas volumetric flow rate in the middle of the specimen amounts to

$$Q_{por} = \frac{2 \cdot Q_{meas} \cdot p_{meas}}{(p_1 + p_2)} = \frac{T_{por}}{T_{meas}} \quad (4)$$

Considering the above, Eq. (2) for the gas permeability coefficient takes the following form.

$$K = \frac{2 \cdot Q_{meas} \cdot p_{meas} \cdot \mu \cdot l}{\pi \cdot \frac{d^2}{4} \cdot (p_2^2 - p_1^2)} \cdot \frac{T_{por}}{T_{meas}} \quad (5)$$

The standard unit from GET 210 implementing the stationary filtration method uses Bronkhorst thermal mass flow meters. These instruments determine the gas mass flow rate to be converted into the volumetric flow rate Q_0 under standard conditions ($p_0 = 0.101325$ MPa and $T_0 = 273.15$ K), constituting the measuring signal of these instruments. Thus, Eq. (5) takes the form.

$$K = 8\frac{Q_0 \cdot p_0 \cdot \mu \cdot l}{\pi \cdot d^2 \cdot (p_2^2 - p_1^2)} \cdot \frac{T_{por}}{T_0} \quad (6)$$

The measurement instruments of the standard unit implementing the stationary filtration method obtain measurement data in the following units: gas volumetric flow rate Q_0 in dm^3/s; gas pressure above p_2 and below p_1 the specimen in MPa; gas temperature in the middle of the specimen in K. The geometric parameters of the specimen are measured using a micrometer (in mm). By substituting these units into Eq. (6) and simplifying the expression, we obtain.

$$K = \frac{\frac{[dm]^3}{[s]} \cdot [MPa] \cdot [\mu Pa] \cdot [s] \cdot [mm]}{[mm]^2 \cdot [MPa]^2} \cdot \frac{[K]}{[K]} = \frac{[dm]^3 \cdot [\mu Pa]}{[mm] \cdot [MPa]}$$
$$= \frac{10^{-3} \cdot [m]^3 \cdot 10^{-6} \cdot [Pa]}{10^{-3} \cdot [m] \cdot 10^6 \cdot [Pa]} = 10^{-12}[m]^2 = [\mu m]^2 \quad (7)$$

Thus, the equation for measuring the gas permeability coefficient K (in 10^{-3} μm^2) at a given pore pressure using GET 210 takes the following form:

$$K = 8000 \cdot \frac{Q \cdot p_0 \cdot \mu \cdot l}{\pi \cdot d^2 \cdot (p_2^2 - p_1^2)} \cdot \frac{T}{T_0}, \quad (8)$$

where K—gas permeability coefficient, 10^{-3} μm^2; Q—volumetric flow rate of gas (dm^3/s) under standard conditions (p_0= 0.101325 MPa and $T_0 = 273.15$ K); l—specimen length, mm; μ – gas viscosity, μPa·s; p_0—standard pressure of 0.101325 MPa; d – specimen diameter, mm; p_2—pressure at the specimen inlet, MPa; p_1—pressure at the specimen outlet, MPa; T – gas temperature during measurement, K; T_0—standard temperature of 273.15 K.

The viscosities of nitrogen and helium are calculated according to Formulas (9) and (10), respectively.

$$\mu_{N_2} = 4.0487 + 0.046105 \cdot T + 0.1606 \cdot P_{por} \tag{9}$$

$$\mu_{He} = 6.2865 + 0.045464 \cdot T + 0.03615 \cdot P_{por} \tag{10}$$

A deviation from Darcy's law can be observed to occur during gas permeability measurements. We can attribute this deviation to the gas slippage phenomenon known as the Klinkenberg effect [3, 11–14]. This effect manifests itself in the dependence of the gas permeability value—as measured and calculated according to Darcy's law—on the pore pressure P_{por} in the specimen. Gas permeability measured at infinite pore pressure is absolute (K_{abs}), approaching that of a liquid not reacting with the rock. Thus, K_{abs} is related to the measured gas permeability at a given pore pressure as follows.

$$K_{abs} = \frac{K}{1 + b/P_{por}}, \tag{11}$$

where K_{abs}—absolute gas permeability coefficient; K—gas permeability coefficient at a given pore pressure; b—Klinkenberg factor depending on the type of rock and filtered gas.

Thus, K_{abs} is determined by measuring gas permeability coefficients at several pore pressure values, constructing a linear regression relationship between the gas permeability coefficient and back pore pressure $K = f(1/p_{por})$, and extrapolating this dependence to the point $1/p_{por} = 0$.

Experimental Studies

Prior to performing the measurements, each specimen was dried at 105 °C in the drying oven for 2 h and then placed in the desiccator for 60 min at room temperature. Next, the geometric parameters of the specimen were measured. To this end, its length l_i and diameter d_i were measured across twenty equally spaced points using the micrometer to calculate the arithmetic means of the specimen's length l and diameter d according to Formula (12):

$$l = \frac{1}{N}\sum_{i=1}^{N} l_i \tag{12}$$

where l—arithmetic mean of the specimen length, mm; l_i—i-th result of measuring the specimen length, mm; N—number of measurements.

The specimen was then placed in a core holder using a rubber seal to create a pressure of ~ 1.5 MPa around the lateral surface of the specimen, avoiding gas slippage between the seal and the specimen. The geometric parameters of the specimen

d and l were then entered into the MG2P500 porosity analyzer software. After that, the gas permeability coefficient was measured three times at the following back pore pressures $1/P_{por}$: (8, 7, 6, 5, 4, 3, 2) MPa^{-1}. During measurements, these values were allowed to deviate by no more than $\pm$ 0.15 MPa^{-1}. Gas permeability coefficients were measured at each pore pressure only after reaching the stationary filtration mode, i.e., the flow rate, pressure, and temperature sensor readings no longer exhibited any trend. Proceeding from a large number of performed measurements, it took a different amount of time for the stationary mode of gas flow rate to set in depending on the absolute gas permeability coefficient K_{abs} when using the standard unit of GET 210: ~30 min for specimens having $K_{abs} = (0.1–1)\ 10^{-3}\ \mu m^2$; ~ 20 min for specimens having $K_{abs} = (1 - 10)\ 10^{-3}\ \mu m^2$; ~15 min for specimens having K_{abs} (10–50) $10^{-3}\ \mu m^2$; ~10 min for specimens having K_{abs} (50–100) $10^{-3}\ \mu m^2$; ~5 min for specimens having K_{abs} over 100 $10^{-3}\ \mu m^2$. Whenever the gas permeability coefficient was measured at a given reverse pore pressure, the software of the standard unit recorded the quantities from Eq. (8) into an MS Excel file. Following three measurements of gas flow rate Q, the pressures above p_2 and below p_1 the specimen, and the gas temperature T at a given back pore pressure, their arithmetic mean values were determined as per Formula (12). The values were then used to calculate the gas permeability coefficient K and the corresponding value of back pore pressure $1/P_{por}$. These operations were performed at each value of back pore pressure 1/Ppor: (8, 7, 6, 5, 4, 3, 2) MPa^{-1}. The obtained gas permeability coefficients were used to construct the linear regression dependence of gas permeability coefficients on back pore pressures $K = f(1/P_{por})$ using the least squares method, followed by extrapolation of this dependence to the point $1/P_{por} = 0$. The gas permeability coefficient at this point was observed to be equal to the absolute gas permeability coefficient K_{abs}. Under favorable conditions, all gas permeability coefficients fitted well onto a straight line $K = f(1/P_{por})$, with the correlation coefficient R^2 of this dependence approaching unity. Whenever any point of the dependence $K = f(1/P_{por})$ fitted poorly onto the straight line, this point was discarded so that the correlation coefficient R^2 remained close to unity and the uncertainty associated with determining the coefficient of absolute gas permeability K_{abs} stayed within acceptable limits. All of the described procedures, starting with the specimen placement in the core holder, were performed using nitrogen and helium.

Determination of the Metrological Characteristics of RMs

As the certified characteristics of RMs, we adopted the gas permeability coefficients measured using helium and nitrogen at given back pore pressures of (8, 7, 6, 5, 4, 3, 2) MPa^{-1}, as well as the absolute gas permeability coefficient.

1. ***Gas permeability coefficients at given back pore pressures***

The certified value of the gas permeability coefficient at a given back pore pressure is calculated according to Eq. (8).

The expanded uncertainty of the certified value at $k = 2$ and $P = 0.95$ is obtained by the formula.

$$U = k \cdot u_c, \tag{13}$$

where k—coverage factor, u_c—combined standard uncertainty of the gas permeability coefficient at a given back pore pressure, 10^{-3} µm^2.

The combined standard uncertainty u_c of the gas permeability coefficient at a given back pore pressure is determined as follows.

$$u_c = \sqrt{u_{char}^2 + u_{stab}^2 + u_{he}^2}, \tag{14}$$

where u_{char}—standard uncertainty associated with RM characterization (for the gas permeability coefficient at a given back pore pressure), 10^{-3} µm^2, u_{stab}—standard uncertainty associated with RM instability, 10^{-3} µm^2; u_{he}—standard uncertainty associated with RM inhomogeneity, 10^{-3} µm^2.

The standard uncertainty associated with RM characterization is calculated by the formula.

$$u_{char} = \sqrt{u_A^2 + u_B^2}, \tag{15}$$

where u_A—Type A standard uncertainty of the gas permeability coefficient at a given back pore pressure, 10^{-3} µm^2, u_B—Type B standard uncertainty of the gas permeability coefficient at a given back pore pressure, 10^{-3} µm^2.

The Type A standard uncertainty u_A of the gas permeability coefficient at a given back pore pressure is obtained according to Formula (16), taking the correlation of input quantities into account. The input quantities x_i and x_j are any two measured input quantities from Eq. (8): volumetric flow rate of gas Q, pressure at the specimen inlet p_2, pressure at the specimen outlet p_1, gas viscosity μ, specimen length l, specimen diameter d, and gas temperature T during measurement.

$$u_A = \sqrt{\sum_{i=1}^{n} c^2(x_i) \cdot u_A^2(x_i) + 2\sum_{i=1}^{n-1}\sum_{j=i+1}^{n} c(x_i) \cdot c(x_j) \cdot u_A(x_i) \cdot u_A(x_j) \cdot r(x_i, x_j)} \tag{16}$$

where $r(x_i,x_j)$—Pearson correlation coefficient; x_i and x_j—input quantities for which the Pearson correlation coefficient is calculated, $u_A(x_i)$ and $u_A(x_j)$—Type A standard uncertainties of input quantities, which are estimated as the standard deviations of the mean values of the input quantities x_i and x_j; $c(x_i)$ and $c(x_j)$—sensitivity coefficients of input quantities x_i and x_j; n—number of all input quantities.

The Pearson correlation coefficient $r(x_i,x_j)$ is determined using the following formula.

$$r(x_i, x_j) = \frac{\sum_{i=1}^{T} (x_{i,k} - \overline{x}_i) \cdot (x_{j,k} - \overline{x}_j)}{\sqrt{\sum_{i=1}^{T} (x_{i,k} - \overline{x}_i)^2 \sum_{i=1}^{T} (x_{j,k} - \overline{x}_j)^2}} \tag{17}$$

where $x_{i,k}$ and $x_{j,k}$—k-th results of measuring input quantities x_i and x_j; $\tilde{x}_{i,k}$ and $\tilde{x}_{j,k}$—arithmetic mean of the measured input quantities x_i and x_j, T—number of measurements for input quantities x_i and x_j.

For example, Eq. (17) takes the following form for the volumetric flow rate of gas Q and the pressure p_2 at the specimen inlet:

$$r(Q, p_2) = \frac{\sum_{i=1}^{T} (Q_k - \overline{Q}) \cdot (p_{2,k} - \overline{p}_2)}{\sqrt{\sum_{i=1}^{T} (Q_k - \overline{Q})^2 \sum_{i=1}^{T} (p_{2,k} - \overline{p}_2)^2}}, \tag{18}$$

where Q_k and $p_{2,k}$—k-th results of measuring the volumetric flow rate Q and pressure at the specimen inlet p_2; $\overline{Q}$ and $\overline{p_2}$—arithmetic means of the measured volumetric flow rate Q and the pressure at the specimen inlet p_2.

The Type A standard uncertainty of the input quantity $u_A(x_i)$, as well as the Type A standard uncertainty of the input quantity $u_A(x_j)$, is calculated using the following formula:

$$u_A(x_i) = \sqrt{\frac{\sum_{i=1}^{T} (x_{i,k} - \overline{x}_i)_k^2}{T(T-1)}}, \tag{19}$$

For example, the Type A standard uncertainty of the volumetric flow rate of gas is determined as follows.

$$u_A(Q) = \sqrt{\frac{\sum_{i=1}^{T} (Q_k - \overline{Q})_k^2}{T(T-1)}}, \tag{20}$$

Formulas given below are used to calculate the sensitivity coefficients for each input quantity c (Q—arithmetic mean of the volumetric flow rate of gas (dm^3/s) under standard conditions (p_0= 0.101325 MPa and T_0 = 273.15 K); l—arithmetic mean of the specimen length, mm; μ—gas viscosity, μPa·s; d—arithmetic mean of the specimen diameter, mm; p_2—arithmetic mean of pressure at the specimen inlet, MPa; p_1—arithmetic mean of pressure at the specimen outlet, MPa; T—arithmetic mean of gas temperature during measurement, K):

$$c(Q) = \frac{\partial K}{\partial Q} = 8000 \cdot \frac{p_0 \cdot \mu \cdot l \cdot T}{\pi \cdot d^2 \cdot \left(p_2^2 - p_1^2\right) \cdot T_0}, \tag{21}$$

$$c(T) = \frac{\partial K}{\partial T} = 8000 \cdot \frac{Q \cdot p_0 \cdot \mu \cdot l}{\pi \cdot d^2 \cdot \left(p_2^2 - p_1^2\right) \cdot T_0}, \tag{22}$$

$$c(l) = \frac{\partial K}{\partial l} = 8000 \cdot \frac{Q \cdot p_0 \cdot \mu \cdot T}{\pi \cdot d^2 \cdot \left(p_2^2 - p_1^2\right) \cdot T_0}, \tag{23}$$

$$c(\mu) = \frac{\partial K}{\partial \mu} = 8000 \cdot \frac{Q \cdot p_0 \cdot l \cdot T}{\pi \cdot d^2 \cdot \left(p_2^2 - p_1^2\right) \cdot T_0}, \tag{24}$$

$$c(d) = \frac{\partial K}{\partial T} = -16000 \cdot \frac{Q \cdot p_0 \cdot \mu \cdot l \cdot T}{\pi \cdot d^3 \cdot \left(p_2^2 - p_1^2\right) \cdot T_0}, \tag{25}$$

$$c(p_2) = \frac{\partial K}{\partial p_2} = -16000 \cdot \frac{p_2 \cdot Q \cdot p_0 \cdot \mu \cdot l \cdot T}{\pi \cdot d^2 \cdot \left(p_2^2 - p_1^2\right)^2 \cdot T_0}, \tag{26}$$

$$c(p_1) = \frac{\partial K}{\partial p_1} = 16000 \cdot \frac{p_1 \cdot Q \cdot p_0 \cdot \mu \cdot l \cdot T}{\pi \cdot d^2 \cdot \left(p_2^2 - p_1^2\right)^2 \cdot T_0}. \tag{27}$$

The Type B standard uncertainty u_B of the gas permeability coefficient at a given back pore pressure is calculated as in Formula (16), taking the correlation between the input quantities into account. However, instead of the standard Type A uncertainties of the input quantities, the standard uncertainties given in the calibration certificates of instruments measuring these input quantities are used.

The absolute gas permeability coefficient K_{abs} is linearly dependent on gas permeability coefficients at a given back pore pressure K. Therefore, the standard uncertainty associated with instability u_{stab} of the gas permeability coefficient at a given pore pressure K is assumed to be equal to that of the absolute gas permeability coefficient K_{abs}. The calculation of uncertainty arising from RM instability u_{stab} is detailed in the next section of this article that deals with determining the uncertainty of the absolute gas permeability coefficient K_{abs}.

The standard uncertainty associated with RM inhomogeneity u_{he} is set to zero since an individual gas permeability coefficient is adopted for each RM at a given back pore pressure, thus avoiding the sources of uncertainty associated with RM inhomogeneity.

2. ***Absolute gas permeability coefficient***

The certified value of the absolute gas permeability coefficient K_{abs} at a given back pore pressure is calculated as follows.

$$K_{abs} = \frac{K_{abs_{He}} + K_{abs_{N_2}}}{2}, \tag{28}$$

where $K_{abs_{N2}}$ and $K_{abs_{He}}$—nitrogen and helium absolute gas permeability coefficients, respectively, 10^{-3} μm^2, with each of them calculated individually taking the

Klinkenberg effect (11) into account by constructing a linear regression relationship between the gas permeability coefficient and back pore pressure $K = f(1/p_{por})$ and extrapolating this dependence to the point $1/P_{por} = 0$. At this point, the gas permeability coefficient is equal to the nitrogen or helium absolute gas permeability coefficient K_{abs}.

The expanded uncertainty of the certified value at $k = 2$ and $P = 0.95$ is obtained according to the formula.

$$U = k \cdot u_c, \tag{29}$$

where k—coverage factor, u_{c}—combined standard uncertainty of the absolute gas permeability coefficient, 10^{-3} μm^2.

The combined standard uncertainty of the absolute gas permeability coefficient u_{c} is determined as follows.

$$u_c = \sqrt{u_{char}^2 + u_{stab}^2 + u_{he}^2}, \tag{30}$$

where u_{char}—standard uncertainty associated with RM characterization (for the absolute gas permeability coefficient), 10^{-3} μm^2, u_{stab}—standard uncertainty associated with RM instability, 10^{-3} μm^2; u_{he} – standard uncertainty associated with RM inhomogeneity, 10^{-3} μm^2.

The standard uncertainty associated with RM characterization is calculated by the formula:

$$u_{char} = \sqrt{\frac{1}{4} \cdot (u_{N_2}^2 + u_{He}^2) + \frac{\Delta^2}{12}} \tag{31}$$

where u_{N_2} and u_{He}—standard uncertainties of the nitrogen and helium absolute gas permeability coefficients, respectively, 10^{-3} μm^2; Δ—difference between the nitrogen and helium absolute gas permeability coefficients, 10^{-3} μm^2.

The standard uncertainties of the nitrogen and helium absolute gas permeability coefficients (u_{N_2} and u_{He}, respectively) are determined via the Monte Carlo method using PTC Mathcad software (Prime 6.0.0.0, Parametric Technology Corporation, 2019) [15]. This procedure consists in generating a large number m (at least 100,000) of dependences $K = f(1/P_{por})$ and calculating the corresponding number of absolute gas permeability coefficients K_{abs}, which are free terms in the linear regression Eq. (32) that is based on the Klinkenberg Eq. (11).

$$K = a \cdot \frac{1}{P_{por}} + K_{abs} \tag{32}$$

where a—constant dependent on the Klinkenberg factor b, MPa.

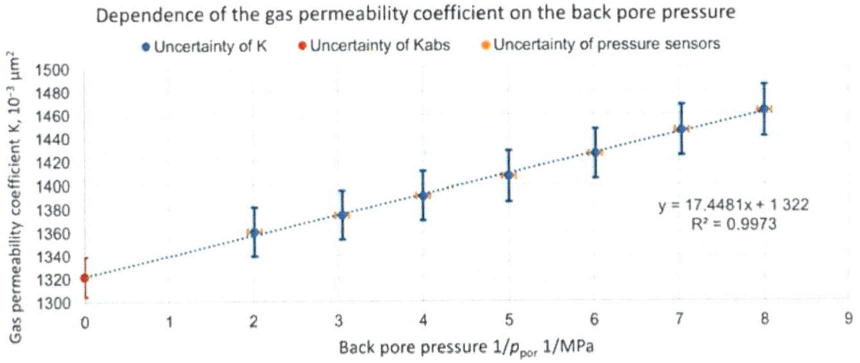

Fig. 1 Dependence of the gas permeability coefficient on the back pore pressure with associated uncertainties

With the known standard uncertainties associated with the characterization of each gas permeability coefficient at given back pore pressures (ordinate axis in Fig. 1) and standard uncertainties of pressure sensors (abscissa axis in Fig. 1), two samples were generated for a series of back pore pressures $1/\ P_{por}$ of (8, 7, 6, 5, 4, 3, 2) MPa^{-1}: a normally distributed sample from m gas permeability coefficients K using the "rnorm" command (PTC Mathcad software) and a sample characterized by uniform distribution from m back pore pressures $1/P_{por}$ using the "runif" command.

The obtained samples of gas permeability coefficients and back pore pressures were used to calculate an array of angular coefficients and free terms in the linear regression Eq. (32). Then, the standard deviation of the array of obtained free terms in the equation (array of absolute gas permeability coefficients K_{abs}) was calculated, which was adopted as the standard uncertainty of the nitrogen absolute gas permeability coefficient u_{N_2}. The same steps were taken to determine the standard uncertainty of the helium absolute gas permeability coefficient u_{He}.

Stability studies were conducted according to GOST ISO Guide 35-2015 [16] using a combination of natural and accelerated aging methods. The results of natural RM aging were supplemented with those of accelerated aging (holding the specimen at an elevated temperature of 105 °C prior to measurement). The characteristics of the absolute gas permeability coefficient were periodically measured at different times, as well as after holding the specimen at an elevated temperature, which simulates longer storage time calculated on the basis of the Van't Hoff equation:

$$\tau = \frac{T_{p.v}}{2^{0.1(t_1 - t_0)}} + K_{abs}, \tag{33}$$

where τ, $T_{\text{p.v}}$—duration of the experiment and the expected period of validity (in days); t_1, t_2—test and storage temperatures of RMs.

A linear model showing the variation of gas permeability characteristics over time is used since there are no other physical reasons to make a different conclusion. The measurement results are processed using regression analysis, as recommended in GOST ISO Guide 35–2015 [16].

The standard uncertainty associated with RM inhomogeneity u_{he} is set to zero since an individual absolute gas permeability coefficient is adopted for each RM, thus avoiding uncertainty sources associated with RM inhomogeneity.

Results and Discussion

The standard uncertainty associated with the instability of RM material is assumed to be equal to the standard deviation arising from the instability of RM material. The results of studying the stability of the absolute gas permeability coefficient are presented in Table 1.

Table 2 presents the metrological characteristics of CRMs calculated according to the previous Sect. ($1/P_{por}$—back pore pressure, MPa^{-1}; K_{N2}—certified value of the nitrogen gas permeability coefficient at a given pore pressure, 10^{-3} μm^2; K_{He}—certified value of the helium gas permeability coefficient at a given pore pressure, 10^{-3} μm^2; K_{abs}—certified value of the absolute gas permeability coefficient, 10^{-3} μm^2; U_o—relative expanded uncertainty of the certified value at $K = 2$ and $P = 0.95$, %).

The conducted studies indicate that the relative expanded uncertainty of the certified RM values within the range of (0.1–5000) 10^{-3} μm^2 does not exceed 3%, which provides from two- to tenfold accuracy margin as compared operating measuring instruments, with an absolute error in measuring the gas permeability coefficient ranging from 6 to 30%. The traceability of the certified values of the developed CRM set to the unit of gas permeability coefficient (μm^2) is provided by using GET 210–2019.

Table 1 Results of studying the stability of the absolute gas permeability coefficient

RM No	Characteristic	Holding time in the thermostat at 105 °C, h			Relative standard uncertainty associated with instability, %
		0	12	15	
		Simulated number of days at 20 °C			
		0	181	226	
1	Absolute gas permeability coefficient, 10^{-3} μm^2	0.6485	0.6525	0.6517	0.96
2		7.736	7.778	7.752	0.97
3		30.16	30.37	30.33	1.05
4		216.81	218.22	217.69	1.06
5		3340	3360	3356	0.94

Table 2 Metrological characteristics of the rock gas permeability CRM set (GSO 11,546–2020/GSO 11,550–2020)

GSO number	GSO 11,546-2020					
$1/P_{por}$, MPa^{-1}	K_{N2}, 10^{-3} μm^2	U_o, %	K_{He}, 10^{-3} μm^2	U_o, %	K_{abs}, 10^{-3} μm^2	U_o, %
2	0.6833	2.3	0.7474	2.2	0.6517	2.6
3	0.6963	2.0	0.7912	2.4		
4	0.7153	2.3	0.8449	2.0		
5	0.7279	2.0	0.8944	2.5		
6	0.7376	2.0	0.9378	2.1		
7	0.7605	2.3	0.9038	2.5		
8	0.7738	2.4	0.8592	2.3		
GSO number	**GSO 11,547-2020**					
$1/P_{por}$, MPa^{-1}	K_{N2}, 10^{-3} μm^2	U_o, %	K_{He}, 10^{-3} μm^2	U_o, %	K_{abs}, 10^{-3} μm^2	U_o, %
2	8.310	2.0	9.666	2.0	7.752	2.7
3	8.621	2.0	10.657	2.2		
4	8.958	2.4	11.598	2.4		
5	9.279	2.0	12.550	2.3		
6	9.617	2.0	13.466	2.4		
7	9.931	2.4	14.396	2.2		
8	10.224	2.3	15.283	2.2		
GSO number	**GSO 11,548–2020**					
$1/P_{por}$, MPa^{-1}	K_{N2}, 10^{-3} μm^2	U_o, %	K_{He}, 10^{-3} μm^2	U_o, %	K_{abs}, 10^{-3} μm^2	U_o, %
2	32.05	3.0	34.55	3.0	30.33	3.0
3	32.73	3.0	36.96	2.9		
4	33.44	3.0	39.07	2.9		
5	34.26	2.9	41.57	3.0		
6	34.95	3.0	43.87	3.0		
7	35.58	3.0	45.93	3.0		
8	35.75	2.9	48.31	2.9		
GSO number	**GSO 11,549–2020**					
$1/P_{por}$, MPa^{-1}	K_{N2}, 10^{-3} μm^2	U_o, %	K_{He}, 10^{-3} μm^2	U_o, %	K_{abs}, 10^{-3} μm^2	U_o, %
2	221.1	2.9	227.6	2.4	217.7	3.0
3	224.3	3.0	232.6	3.0		
4	225.8	3.0	237.7	3.0		
5	227.4	3.0	243.7	3.0		
6	229.3	3.0	248.8	3.0		
7	231.0	3.0	254.0	3.0		
8	233.0	2.8	265.3	2.9		

(continued)

Table 2 (continued)

GSO number	GSO 11,550–2020					
$1/P_{por}$, MPa^{-1}	K_{N2}, 10^{-3} μm^2	U_o, %	K_{He}, 10^{-3} μm^2	U_o, %	K_{abs}, 10^{-3} μm^2	U_o, %
2	3334	2.1	3405	2.4	3356	2.5
3	3392	2.1	3429	2.3		
4	3413	2.3	3454	2.1		
5	3435	2.3	3471	2.3		
6	3451	2.4	3493	2.2		
7	3463	2.3	3514	2.3		
8	3471	2.3	3553	2.4		

Conclusion

As a result of the study, a set of rock gas permeability RMs (imitators) was approved (GSO 11,546–2020 / GSO 11,550–2020). The set includes five cylinders made of molded sintered refractory material (Al_2O_3 content of at least 85%), each measuring 30 mm in height and outer diameter. Each RM is engraved with an index, with the whole set placed in a labeled wooden case. The period of RM validity is two years. These are the first type-approved RMs with traceability to the GET 210–2019, ensuring metrological traceability and the reliability of gas permeability coefficient measurements, which in turn leads to higher measurement quality and finds practical applications at petrophysical research laboratories.

One possible approach to further improving GSO 11,546–2020/GSO 11,550–2020 can be to create imitators carrying sintered alumina ceramics in a hollow stainless-steel cylinder. The surface of such RMs is more durable, as well as characterized by greater smoothness than that of ceramics.

Acknowledgements We express our gratitude to Andrey Gnedov (Deputy Director General of EcogeosProm) for discussing the study results pertaining to the development of the certified RM set.

Author Contributions Both authors have contributed equally to the work.

Conflict of Interest Egor Sobina is one of the editors of the book *Reference Materials in Measurement and Technology.*[12]

The article was prepared on the basis of a report presented at the IV International Scientific Conference "Reference Materials in Measurement and Technology" (St. Petersburg, December 1–3, 2020). The article was admitted for publication after the abstract was revised, the article was formalized and the review procedure was carried out.

The version in the Russian language is published in the journal "Measurement Standards. Reference Materials" 2021;17(2):5–18. https://doi.org/10.20915/2687-0886-2021-17-2-5-18.

[12] Medvedevskikh S., Sobina E., Kremleva O., and Okrepilov M. (eds.), Reference Materials in Measurement and Technology. RMMT, Springer, Cham., Switzerland (2020).

References

1. Sobina EP (2016) Development of certified reference materials set for opened porosity of solid substances and materials (imitators). Refer Mater (2):36–43. https://doi.org/10.20915/2077-1177-2016-0-2-36-43 (In Russ.)
2. Kobranova VN (1986) Petrophysics: textbook allowance, 2nd., rev. and add. Nedra, Moscow, 272p (in Russ.)
3. Troitskiy VM (2017) Phenomenological approach to analysis of empirical data on gas permeability in porous media. A true reason for Klinkenberg's effect. Vesti gazovoy nauki 2(30):110–124 (in Russ.)
4. Kozyryatsky NG (2016) Standard core samples for the metrological support to measure absolute gas permeability in the core. Karotazhnik 9(267):73–84 (in Russ.)
5. GOST 26450.2–85 (1985) Rocks. Method for determination of absolute gas permeability coefficient by stationary and non-stationary filtration. Izdatel'stvo standartov, Moscow (in Russ)
6. Peskov AV (2020) Features of measuring absolute permeability of rocks. Vestnik of Samara State Technical University (Technical Sciences Series) 28(2):73–83. https://doi.org/10.14498/tech.2020.2.5 (in Russ)
7. ASTM D 4525–90 (2001) Standard test method for permeability of rocks by flowing air. https://www.astm.org/DATABASE.CART/HISTORICAL/D4525-90R01.htm
8. RP 40 Recommended practices for core analysis. http://w3.energistics.org/RP40/rp40.pdf
9. Sobina EP (2020) State primary standard of units of specific adsorption of gases, specific surface area, specific pore volume, pore size, open porosity and gas permeability coefficient of solids and materials GET 210–2019. Measur Techniques. (12):3–12. https://doi.org/10.32446/0368-1025it.2020-12-3-12
10. GOST 34100.3—2017/ISO/IEC Guide 98–3:2008 (2018) Uncertainty of measurement — Part 3: Guide to the expression of uncertainty in measurement. Standartinform, Moscow (In Russ)
11. Klinkenberg LJ (1941) The permeability of porous media to liquids and gases. Drilling and Production Practice pp. 200–211
12. Kazemi M, Takbiri-Borujeni A (2015) An analytical model for shale gas permeability. Int J Coal Geol 146:188–197
13. Wang S, Lukyanov AA, Wu YS (2019) Second-order gas slippage model for the Klinkenberg effect of multicomponent gas at finite Knudsen numbers up to 1. Fuel 235:1275–1286. https://doi.org/10.1016/j.fuel.2018.08.113
14. Okwananke A, Hassanpouryouzband A, Vasheghani M, Yang J, Tohidi B, Chuvilin E, Istomin V, Bukhanov B (2019) Methane recovery from gas hydrate-bearing sediments: an experimental study on the gas permeation characteristics under varying pressure. J Petrol Sci Eng 180:435–444. https://doi.org/10.1016/j.petrol.2019.05.060
15. Parametric Technology Corporation (2019) Mathcad Prime 6.0.0.0 [software]
16. GOST ISO Guide 35-2015 (2016) Standartnye obraztsy. Obshchie i statisticheskie printsipy sertifikatsii (attestatsii). Standartinform, Moscow (in Russ)

Ilya P. Aronov Junior Researcher at the Laboratory for Metrological Support of Nano Industry, Spectral Analysis Methods and Reference Materials, UNIIM – Affiliated Branch of the D. I. Mendeleyev Institute for Metrology.

Egor P. Sobina Dr. Sci. (Eng.), Director, UNIIM – Affiliated Branch of the D. I. Mendeleyev Institute for Metrology

Luminescence Measurements: Reference Materials for Composition of Sodium Adenosine Triphosphate Aqueous Solution

Marina M. Chugunova, Natalia Yu. Gryazskikh, Irina N. Zyablikova, Alexander V. Ivanov, and Anna N. Shobina

Abstract This article evaluates the possibility of using luminometry as one of the most effective methods of rapid microbiology in metrological support during monitoring the level of sanitation and hygiene in industry. The key stages in the development of a certified reference material (CRM) for the composition of sodium adenosine triphosphate (sodium ATP) are described, which included the analysis of similar CRMs, selection of starting materials, experimental studies, and establishment of the metrological characteristics of the developed CRM traceable to the State Primary Measurement Standard. The CRM stability was studied by spectrophotometry, considering that the selective optical density at the characteristic wavelength is directly proportional to the concentration of the absorbing component (sodium ATP). The uncertainty of measurement results was evaluated by the isochronous method. The study has confirmed the possibility of ensuring the traceability of luminescence measurements to the State Primary Measurement Standard by the sodium ATP concentration. An optimal value of the sodium ATP concentration was selected, and the metrological characteristics of GSO 11,606–2020 CRM for the composition of sodium ATP aqueous solution were established (the mass concentration of 0.09–0.11 g/dm^3 and the extended uncertainty of 0.015 g/dm^3). The developed CRM can be recommended for performing tests and calibrations when verifying luminescence measuring instruments for providing an unbiased estimation of their measuring capabilities and the associated measurement errors.

M. M. Chugunova (✉) · N. Yu. Gryazskikh · I. N. Zyablikova · A. V. Ivanov · A. N. Shobina
All-Russian Research Institute for Optical and Physical Measurements (VNIIOFI), Ozernaya st., 46, Moscow 119361, Russia
e-mail: chugunova@vniiofi.ru

N. Yu. Gryazskikh
e-mail: griazskih@vniiofi.ru

I. N. Zyablikova
e-mail: center_gso@vniiofi.ru

A. V. Ivanov
e-mail: ivanov@vniiofi.ru

A. N. Shobina
e-mail: shobina@vniiofi.ru

S. V. Medvedevskikh et al. (eds.), *Reference Materials in Measurement and Technology*, https://doi.org/10.1007/978-3-031-06285-8_3

Keywords Reference material · Sodium adenosine triphosphate · Luminescence · Chemiluminometer · Bioluminescence · Metrology

Introduction

Food manufacturers all over the world aim to continuously improve the quality and shelf life of food products. A necessary condition for achieving these objectives is the high level of sanitation and hygiene at all production sites and facilities. According to the requirements of the Technical Regulations of the Customs Union TP TC 021/2011 [1], starting from February 2015, all food manufacturers should implement the Hazard Analysis and Critical Control Points (HACCP) management system [2]. HACCP is widely used for ensuring the safety of food processing and mitigating the risks involved with production and marketing processes.

Golic et al. [3] reviewed the results of applying standard evaluation methods[1] for determining of the rate of biological and other contamination of working surfaces, water, and personnel hands. The researchers showed that a serious drawback of standard methods is a delay in obtaining the evaluation results. Moreover, additional control of the residual products after washing and disinfecting the equipment required by HACCP extends the entire validation process. We believe that luminometry, as a method of "rapid microbiology" [4], is an optimal solution to providing efficient quality management that ensures a high level of sanitation and hygiene at reduced time costs. As an example, Table 1 compares different methods in term of the time required for determining the rate of biological contamination.

The operational principle of luminometers is based on the effect of bioluminescence, i.e. the ability of the working substance of a test system in combination with the products of microorganism vital activity to release luminous energy during particular chemical reactions.

Table 1 Comparative analysis of the time required for determining the rate of biological contamination by the methods of direct microbiology, real-time PCR, and luminometry

Measurement procedure	Time of obtaining the results
Direct microbiology	6–30 h
Real-time PCR	1.5 h
Luminometry	15 s

[1] For the microbiological examination of swab samples, the following standards were used:

– BAS EN ISO 4833:2006 (Microbiology of food and animal feeding stuffs, 2006) for determining the number of microorganisms.

– BAS ISO 21528–2:2008 (Microbiology food and animal feeding stuffs, 2013) for determining the number of enterobacteria.

– BAS EN ISO 11290–1/A1:2005 (Microbiology of food and animal feeding stuffs, 2005) for the detection of Listeria monocytogenes.

– BAS EN ISO 6579/Cor2:2010 (Microbiology of food and animal feeding stuffs, 2010) for the detection of Salmonella spp.

One approach to determining contaminants in a test sample consists in assessing the integral indicator of contamination by the intensity of bioluminescent emission [5–9]. In this case, the quantitatively indicated values are inversely proportional: i.e., the higher the pollution, the lower the intensity of the luminescent signal, since contaminants inhibit enzymatic reactions in living organisms.

An alternative approach to identifying contaminants using bioluminescence [10] is an analysis in which the intensity of the recorded signal is directly proportional to the concentration of microorganisms or their residues. The possibility of such an analytical approach is associated with the fact all herbal- and animal-based materials (including particles of the corresponding nutritional products), living and dead bacteria, fungal and other microorganisms comprise adenosine triphosphate (ATP) molecules [10]. During the analysis, the ATP molecule interacts with the working substances (luciferin and luciferase), producing an emission proportional to the available amount of ATP. Although luciferins and luciferases present in different organisms differ in their chemical structure, all chemiluminescent (bioluminescent) reactions require molecular oxygen and yield intermediate complexes—organic peroxide compounds. In the course of the reaction, these complexes dissociate, releasing photon energy and exciting the molecules responsible for light emission:

$$E + LH_2 + ATP \rightarrow E - LH_2\text{—}AMP + PP$$

$$E\text{—}LH_2\text{—}AMP + O_2 \rightarrow LO*\text{—}E\text{—}AMP + CO_2$$

$$LO*\text{—}E\text{—}AMP \rightarrow E + LO + AMP + photon$$

where ATP-adenosine triphosphate; AMP-adenosine monophosphate; PP-pyrophosphate; E-luciferase; LH_2-luciferin; LO* and LO-the reaction product (oxyluciferin) in the excited and ground states, respectively; LO*-E-AMP-a transition-state intermediate.

The quantum yield of bioluminescent reactions is extremely high (from 10% up to 100%). Such a significant efficiency of the process is achieved due to the involvement of a highly specific protein biocatalyst-luciferase. It was experimentally demonstrated [11, 12] that the rate of biological contamination can be determined by measuring the intensity of the emission produced by bioluminescent reactions. The greater the emission intensity, the more microorganisms are present in a particular site. In the absence of ATP, i.e., in the absence of microorganisms, bioluminescence is never observed.

Luminometers display the intensity of the light emitted by a sample in relative luminescence units (RLUs). Although RLUs are not an absolute physical parameter, these units allow the intensity of bioluminescence to be assessed objectively, depending on ATP concentration. Thus, for SystemSURE Plus devices manufactured by Hygiena International Ltd. (Great Britain), 1 RLU of bioluminescence equals approximately 1 fmol (10^{-15} mol) of ATP. According to [13], such an amount of intracellular ATP is contained in several microbial cells, which is equivalent to a single colony-forming unit (CFU) on a nutrient medium. However, commercial devices produced by different manufactures are calibrated differently. As a result, luminometers might show different RLU values for samples having the same level of surface contamination, thus requiring additional procedures for converting the obtained values into the corresponding amount of ATP.

The metrological traceability of luminescence devices is established based on calibration graphs, which correlate the bioluminescence intensity expressed in RLU and the concentration of ATP in a solution. Our analysis shows that the Federal Information Fund for Ensuring the Uniformity of Measurements contains dozens of certified luminometers and analyzers, which are validated using certified ATP solutions.

Our analysis of chemical or biological luminescence-based measuring instruments included in the Federal Information Fund for Ensuring the Uniformity of Measurements revealed that, in Russia, there are no certified values or certified reference materials necessary for ensuring the metrological traceability of luminometers. Therefore, our aim was to develop a reference material (RM) for sodium adenosine triphosphate aqueous solution (sodium ATP), which could be used to ensure the metrological traceability of luminescence measuring devices to a unit of mass concentration.

This article describes key stages in the development of an RM with the certified value of sodium ATP, including: (1) the review of analogous RMs; (2) the selection of starting materials; (3) experimental studies; and (4) the establishment of the RM metrological characteristics traceable to the State Primary Standard of Units of Mass (Molar) Fraction and Mass (Molar) Concentration of Components in Liquid and Solid Substances and Materials based on Spectral Methods SPS 196–2015.[2]

Materials and Methods

Starting Materials

Prior to RM development, we analyzed available RMs for sodium ATP with various certified characteristics. As a result, no RMs (CRMs) with the certified value of ATP concentration were found in the RF Information Fund for Ensuring the Uniformity of Measurements. A similar analysis of RMs produced by such foreign metrological institutes as NIST (USA) and BAM (Germany) also revealed the absence of RMs with the certified characteristics of ATP concentration. Conversely, a large selection of ATP sodium salts of various purity is presented on the Sigma-Aldrich Production

[2] GET 196–2015 State Primary Standard of Units of Mass (Molar) Fraction and Mass (Molar) Concentration of Components in Liquid and Solid Substances and Materials based on Spectral Methods. In Federal Information Fund for Ensuring the Uniformity of Measurements. https://fgis.gost.ru/fundmetrology/registry/12/items/397882.

Table 2 Sodium ATP of various purity produced by Merck (formerly Sigma-Aldrich production GmbH)

Product No	Product name	Product objective
FLAAS	Adenosine 5′-triphosphate (ATP) disodium salt hydrate (lyophilized powder, ~ 1 mg ATP)	ATP measurement by luminescence
A26209	Adenosine 5′-triphosphate disodium salt hydrate, 99%	ATP product serves as a coenzyme in a wide range of enzymatic reactions, and as a substrate for ATP-dependent enzymatic systems
A2383	Adenosine 5′-triphosphate disodium salt hydrate, ≥ 99%	The product is used for preparing ATP reference solutions for determining the ATP level in bacterial cultures

GmbH website (Table 2) manufactured by Merck (Germany), one of the largest manufacturers of high-technology materials for biopharmaceutics and biosciences. The materials listed in Table 2 are used, inter alia, for bioluminescent studies.

However, due to the high cost of this materials, we decided to apply a water-soluble reagent produced by Xi'an Lyphar Biotech Co., Ltd. (China) with a purity comparable with the analogues (the mass fraction of the main component is not less than 99.3%) as a starting material for RM preparation.

Equipment and Experimental Studies

RMs for the composition of sodium ATP aqueous solution were prepared by dissolving the starting material in distilled water. To that end, the following equipment was used: laboratory balances Sartorius ME36S, class of special accuracy (I) according to OIML R 76-1-2014; a graduated bulb of 1st accuracy class according to GOST 1770; and a distiller D-4 TU 64–1-1640–72.

The purity of the starting material was determined by a chromate-mass-spectrometer 320-MS (Varian B.V., Netherlands) included in SPS 196–2015. The chromatogram of the starting material (Fig. 1) in the water-soluble reagent under study revealed no impurities; therefore, the mass fraction of the main component (sodium ATP) of 99.3% was taken as the purity of the starting material.

Since the selective optical density at the characteristic wavelength is directly proportional to the concentration of the absorbing component (sodium ATP), we decided to apply the method of spectrometry. This method measures the concentration of dissolved substances by the amount of light absorbed by the solution. The optical density of sodium ATP solution in transmitted light was measured at a characteristic wavelength of 258 nm (Fig. 2).

In order to ensure the independence of measurements, the certified values of the mass concentration of sodium ATP were determined on the basis of calibration. To

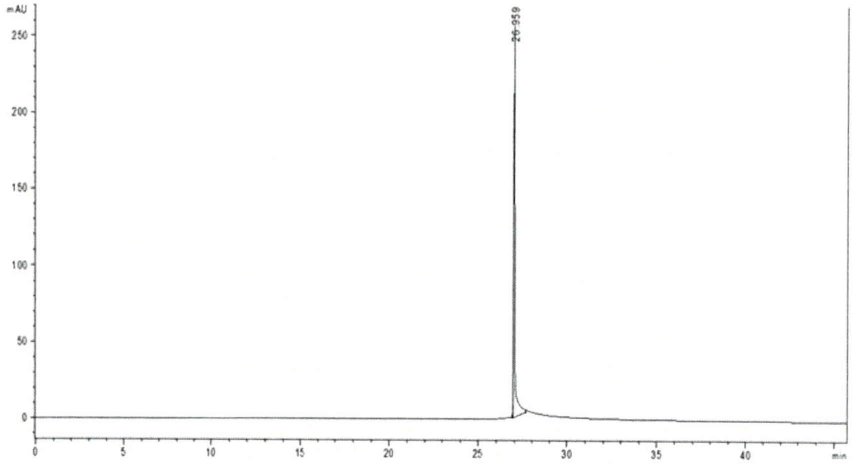

Fig. 1 Chromatogram of the starting reagent produced by Xi'an Biotechnology Company Lyphar, LTD

Fig. 2 Absorption spectrum of the studied sodium ATP solutions in transmitted light

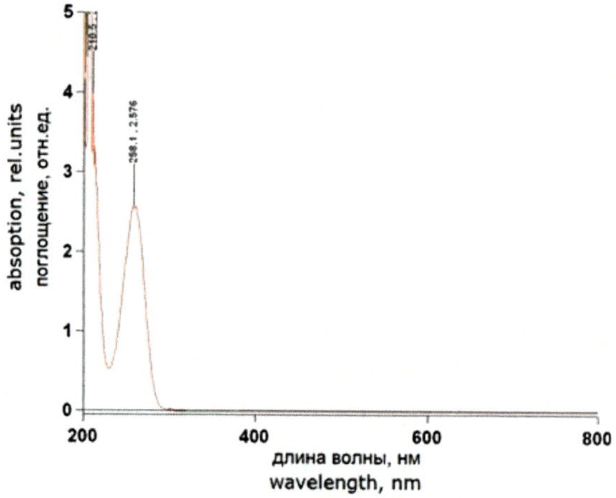

that end, a powder of pure sodium ATP produced by MP Biomedicals, LLC was selected as a basis for calibration solutions (the mass fraction of the main component as defined by the manufacturer was 99.2%).

The optical density of the solutions was measured by spectrophotometry using a Cary 50 Spectrophotometer (Varian Ltd., Australia) included in SPS 196–2015 with the following metrological characteristics:

- a spectral range of 190–1100 nm;
- a measurement range of the directional transmission spectral coefficients of 0–100%;
- the permissible absolute error when determining the wavelength of ± 1.0 nm.

CRM Stability

A study into the stability of the developed RMs was carried out in accordance with RMG 93–2015 [14]. To that end, all RM samples were divided into two lots and stored for 30 days: one lot at a temperature of (5 ± 3) °C, and another lot-at a temperature of 40 °C (which corresponds to one year of storage at a temperature of 5 °C). During the storage period, seven pairs of measurements were carried out under repeatability conditions.

For solutions with a concentration of about 0.1 g/dm^3 and higher, no statistically significant changes were observed during the research period. For samples with a lower concentration, changes in the optical characteristics, indicating their irreversible changes, were observed. Since samples with a minimum stable concentration are required for metrological application, the value of 0.1 g/dm^3 was selected as a basic value for the CRM.

Uncertainty of the CRM Certified Value

The total standard uncertainty of the CRM certified value was determined by the formula:

$$u_c = \sqrt{u_{char}^2 + u_{stab}^2 + u_M^2}, \tag{1}$$

where u_{char}-standard uncertainty associated with the method of determining the RM certified value;

u_{stab}-standard uncertainty associated with the RM stability;

u_M-standard uncertainty associated with the purity of starting materials (according to the manufacturer $u_M = 1\%$).

The extended uncertainty U was calculated based on the total standard uncertainty u_c and the coverage coefficient k = 2 according to the formula:

$$U = ku_c$$

Table 3 presents the values of the mass concentration of ATP and their extended uncertainty obtained taking into account the stability of solutions and the error of the method.

Table 3 Metrological characteristics of the developed CRM

CRM title	Certified value (mass concentration), g/dm^3	Extended uncertainty (if k = 2), g/dm^3
Reference material for the composition of sodium adenosine triphosphate aqueous solution[i]	0.09–0.11	0.015

[i]The RM producer is the All-Russian Scientific Research Institute for Optical and Physical Measurements (VNIIOFI), Moscow, Russia

Results and Discussion

In this work, we investigated the physical and metrological characteristics of aqueous solutions of sodium ATP with the purpose of developing a respective CRM. The starting material was a reagent produced by Xi'an Lyphar Biotech Co., Ltd., the purity of which was tested by chromate-mass-spectrometry. The stability of solutions was tested using the spectrophotometry method.

The conducted research confirmed the possibility of using the sodium ATP concentration for ensuring the metrological traceability of luminescence measurement results to SPS 196–2015. The optimal value of the ATP mass concentration was established, along with the metrological characteristics of the developed CRM: the sodium ATP mass concentration of 0.09–0.11 g/dm^3 and the extended uncertainty of 0.015 g/dm^3. By dissolving the initial CRM following the recommendations of the producer, sodium ATP concentrations of up to 0.000001 g/dm^3 can be obtained. These concentrations correspond to the substance amount of 10 fmol for a dosage of 5 μl, thus meeting the detection limit of current measuring instruments. The developed CRM has no analogues on the Russian market.

Conclusion

Until recently, ensuring the metrological traceability of luminescence measuring instruments to mass concentration units has been hampered by the lack of appropriate CRMs. As a result of our research, the GSO 11606–2020 CRM for the composition of sodium ATP aqueous solution composition was developed. The CRM metrological characteristics were established, including the certified value of sodium ATP mass concentration (0.09–0.11 g/dm^3) and the extended uncertainty (0.015 g/dm^3). The developed CRM is the first reference material traceable to the State Primary Standard SPS 196–2015, which can be used for ensuring the metrological traceability and validity of measurement results when verifying, calibrating, and testing measuring instruments (chemiluminometers, bioanalyzers, etc.). The CRM allows an unbiased estimation of the measuring capability of luminescence instruments, including the determination of measurement errors. The developed CRM can be used to ensure the uniformity of measurements in the field of ATP-luminometry for storing, reproducing, and transferring the unit of mass concentration.

Future research focused on improving GSO 11606–2020 may address the development of a system for an accurate express dilution of the initial CRM solution to lower concentrations immediately prior to measurements at the user's facility.

Acknowledgements The authors express their gratitude to M. Y. Larionov and A. O. Elenskii for their assistance in the supply of starting materials for the research. All measurements were carried out using the VNIIOFI equipment.

Author Contributions M. M. Chugunova: research concept development, supervision of the process of CRM certification during interlaboratory comparison, obtaining and processing experimental data.

N. Yu. Gryazskikh: research concept development, critical data analysis, manuscript revision.

I. N. Zyablikova: performance of calculations, preparing documents for CRM certification, packaging.

A. V. Ivanov: research concept development, supervision of the process of CRM certification during interlaboratory comparison.

A. N. Shobina: preparation of technical specifications, literature review, critical data analysis, manuscript revision.

Conflict of Interest The article was prepared on the basis of a report presented at the IV International Scientific Conference "Reference Materials in Measurement and Technology" (St. Petersburg, December 1–3, 2020). The article was admitted for publication after the abstract was revised, the article was formalized and the review procedure was carried out.

The version in the Russian language is published in the journal "Measurement Standards. Reference Materials" 2021;17(3):35–44. https://doi.org/10.20915/2687-0886-2021-17-3-35-44.

References

1. TR CU 021/2011 Technical regulations of the customs union "On food safety". https://legalacts.ru/doc/reshenie-komissii-tamozhennogo-soiuza-ot-09122011-n-880-o/ (in Russ)
2. Karen L, Schlosser W (2002) Hazard analysis and critical control point (HACCP) history and conceptual overview. Risk Anal 3:547–552. https://doi.org/10.1111/0272-4332.00038
3. Golic B, ILIĆ T, Kalaba V, GOLIĆ M (2019) Microbiological purity testing in food production and marketing. Veter J Repub Srpsk 19(1). https://doi.org/10.7251/VETJEN1901064G
4. Osimani A, Aquilanti L, Clementi F (2014) The luminometer as a useful tool in the assessment of surface cleanliness: the case study of a university canteen. Indus Alimen 53(545):12–18
5. Baigina EM, Rimatskaya NV, Stepanova LV, Kratasyuk VA (2017) On the possibility of application of bioluminescent enzyme system for the analysis of soil contamination of the Krasnoyarsk territory. *Izvestiya Irkutskogo gosudarstvennogo universiteta. Seriya «Biologiya. Ecologiya»* 21:21–30 (in Russ)
6. Morozova EV (2017) The use of biotesting as an integral method for assessing the toxicity of environmental factors. Trends in the development of science and education (26–3):9–12 (in Russ) https://doi.org/10.18411/lj-31-05-2017-39
7. Simbekova EN, Kondik AM, Kratasyuk VA (2013) Bioluminescent enzymatic rapid assay of water integral toxicity. Environ Monit Assess 185(7):5909–5916. https://doi.org/10.1007/s10661-012-2994-1

8. Abraham JS et al (2019) Soil ciliates of the Indian Delhi Region: Their community characteristics with emphasis on their ecological implications as sensitive bio-indicators for soil quality. Saudi J Biol Sci 26(6):1305–1313. https://doi.org/10.1016/j.sjbs.2019.04.013
9. Kim SB (2016) Bioluminescence: methods and protocols. Methods Molec Biol Book Series 1461. https://doi.org/10.1007/978-1-4939-3813-1
10. Gayford C (1986) Adenosine triphosphate. In: *Energy and Cells: Dimensions of Science*. Palgrave, London https://doi.org/10.1007/978-1-349-08159-2_2
11. Sezemin IA (2007) Use of express train-method adenosine triphosphate-luminometry at sanitary-and-hygienic researches in food processing industry. Food Ind 3:18–19 (in Russ)
12. Fordet SR et al (1996) Use of firefly luciferase for ATP measurement: other nucleotides enhance turnover. Biolumin Chem 11(3):149–167. https://doi.org/10.1002/(SICI)1099-1271(199605)11:3<3c149::ПОМОЩЬ-BIO411>3e3.0.CO;2-Q
13. Zverev DS (2008) Determination of the number of living microbial cells in anti-brucellosis vaccines by the bioluminescent method: Dissertation of the candidate of biological sciences. Moscow, p 23 (in Russ)
14. RMG 93–2015 GSI. State system for ensuring the uniformity of measurements. Estimation of metrological characteristics of reference materials. Standartinform, Moscow, 2011, p 30 (in Russ)

Marina M. Chugunova Leading Researcher, the All-Russian Research Institute for Optical and Physical Measurements.

Natalia Yu. Gryazskikh Head of Sector, the All-Russian Research Institute for Optical and Physical Measurements.

Irina N. Zyablikova Engineer, the All-Russian Research Institute for Optical and Physical Measurements.

Alexander V. Ivanov Head of Testing and Certification Department, the All-Russian Research Institute for Optical and Physical Measurements.

Anna N. Shobina Head of Testing Department, the All-Russian Research Institute for Optical and Physical Measurements.

Identification of Reference Materials of Azithromycin Composition Using Spectral Methods

Vladimir I. Gegechkori, Nikolai A. Shulga, Olga Yu. Shchepochkina, Neonila V. Gorokhovets, Anatolii A. Levko, Nataliia N. Chadova, and Vladislav N. Shestakov

Abstract Standardization is essential in the process of developing novel drugs, as well as updating approaches to the quality control of medicines in circulation. Establishing methods of quality control using reliable, accurate, and sensitive physicochemical methods is a crucial issue in standardization. This, in turn, requires the development of reliable reference materials (RMs). This article considers the development of methods for determining the structure of azithromycin, which is important for ensuring the purity of RMs and selecting an adequate technique of quantification according to the current regulatory documents, as well as the State Pharmacopoeia of the Russian Federation (XIV Edition).

Keywords Azithromycin · Reference materials · Physicochemical methods · Standardization · NMR spectroscopy · IR spectroscopy · Mass spectrometry

Introduction

Physical and chemical analysis techniques are increasingly used in developing approaches to drug quality assessment. Their use requires reference materials (RMs) of an active pharmaceutical ingredient (API), intermediates, inactive ingredients, and impurities. Furthermore, RMs are used in related scientific fields, which are of great practical importance for Pharmacia. These fields include fundamental and applied research of pharmacokinetics of drugs, therapeutic drug monitoring for further correction of medicinal treatment, chemical and toxicological analysis for

V. I. Gegechkori (✉) · N. A. Shulga · O. Yu. Shchepochkina · N. V. Gorokhovets
I.M. Sechenov First Moscow State Medical University (Sechenov University), 121019 Moscow, Russia

O. Yu. Shchepochkina
e-mail: shchepochkina_o_yu@staff.sechenov.ru

N. V. Gorokhovets
e-mail: gorokhovets_n_v@staff.sechenov.ru

A. A. Levko · N. N. Chadova · V. N. Shestakov
State Institute of Drugs and Good Practices, Moscow, Russia

S. V. Medvedevskikh et al. (eds.), *Reference Materials in Measurement and Technology*,
https://doi.org/10.1007/978-3-031-06285-8_4

determining further detoxification therapy, or death from a particular drug. Thus, establishing a system of circulation and control over RMs of different levels within the territory of the Russian Federation is a topical issue for modern science and the pharmaceutical industry [1, 2].

This study aims to develop a method for analyzing azithromycin samples when preparing RMs of azithromycin composition, which can be further used, e.g., to identify and quantify this drug in various pharmaceutical dosage forms. Approaches to selecting quality parameters and methods of their determination should be strictly controlled following the structural and physicochemical characteristics of an RM and its area of application of [3, 4].

Azithromycin ((2R, 3S, 4R, 5R, 8R, 10R, 11R, 12S, 13S, 14R) -3,4,10-Trihydroxy-13 - [(2,6-dideoxy-3-C-methyl-3- O-methyl-α-L-ribo-hexopyranosyl) oxy] -3,5,6,8,10,12,14-heptamethyl-11 - {[3,4,6-trideoxy-3- (dimethylamino) -β -D-xylo-hexopyranosyl] oxy} -2-ethyl-1-oxa-6-azacyclopentadecan-15-one dihydrate) is a representative of the macrolide group, in particular, azalides and has a wide spectrum of antimicrobial activity. The quality requirements of any antibiotic should remain high at any stage of the life cycle due to their hepatotoxicity characteristics and serious potential side effects [5]. The chemical structure of the investigated compound is shown in Fig. 1. The molecular weight of azithromycin is M = 748.5 [6]

Approaches to the development of RMs are described in many international and domestic regulatory documents. However, they all imply the presence of such quality parameters of the finished product as *Authenticity*, *Purity*, and *Quantitative content* [7–11].

One of the most important quality characteristics of an RM is *Authenticity*. IR spectrometry is conventionally used for its determination; however, it is advisable to use several methods when developing drug RMs [12–14]. These include, for example, UV spectrophotometry, mass spectrometry, elemental analysis, and several other techniques, the combination of which ensures an accurate characterization of the substance. An attempt to classify RMs and describe their preparation process was made in an ISO report (ISO/TR 10,989: 2009 E) and in the current Pharmacopoeia of the Russian Federation, XIV Edition [8, 16]. The International Council

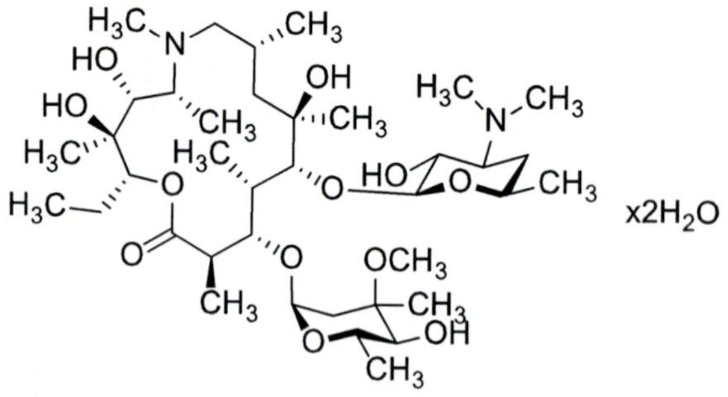

Fig. 1 Chemical structure of azithromycin

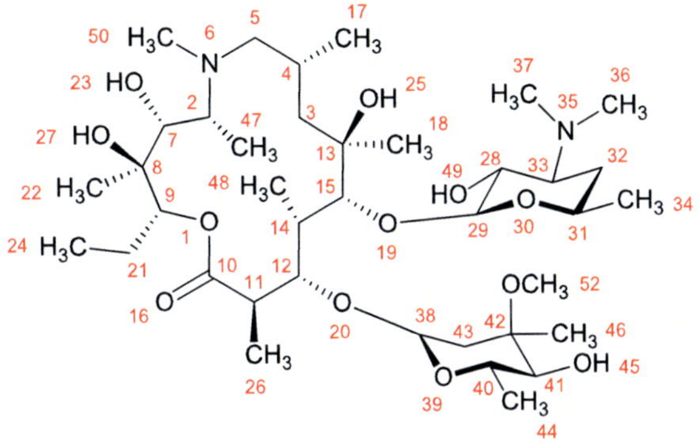

Fig. 2 Chemical structure of azithromycin with numbered atoms

for Harmonization of Technical Requirements for Pharmaceuticals for Human Use (ICH) harmonizes the instructions and requirements for pharmaceutical registration.

In this article, we review the above documents and guidelines published by the ICH with a particular focus on IR spectrometry, NMR spectroscopy, and MS spectrometry as mandatory and independent methods for determining the structure of obtained samples [15, 16].

Materials and Methods

We investigated azithromycin samples from 3 different experimental series – Series 1, Series 2, and Series 3, which were initially controlled according to the current regulatory documents on the development of RMs for active pharmaceutical substances.

The infrared spectrum of azithromycin (Fig. 2) was recorded in the range from 4000 to 400 cm^{-1} using a Bruker Alpha FT-IR spectrophotometer. To that end, a potassium bromide disk (1 - 2 mg of substance in 300 mg of potassium bromide) was prepared.

NMR Spectroscopy

Chemical shifts of azithromycin substances were determined by the ^{1}H and ^{13}C nuclear magnetic resonance (NMR) spectroscopy[1] [8]. ^{1}H and ^{13}C NMR spectra (Figs. 3 and 4, respectively) of a sample solution in deuterated dimethyl sulfoxide (DMSO-d6) were recorded using a Bruker AVANCE III 400 MHz UltraShield Plus spectrometer with operating frequencies 400 MHz and 100 MHz, respectively.

[1] RF, General Pharmaceutical Norms and Regulations GPNR.1.2.1.2.0007.15 "Nuclear magnetic resonance spectroscopy".

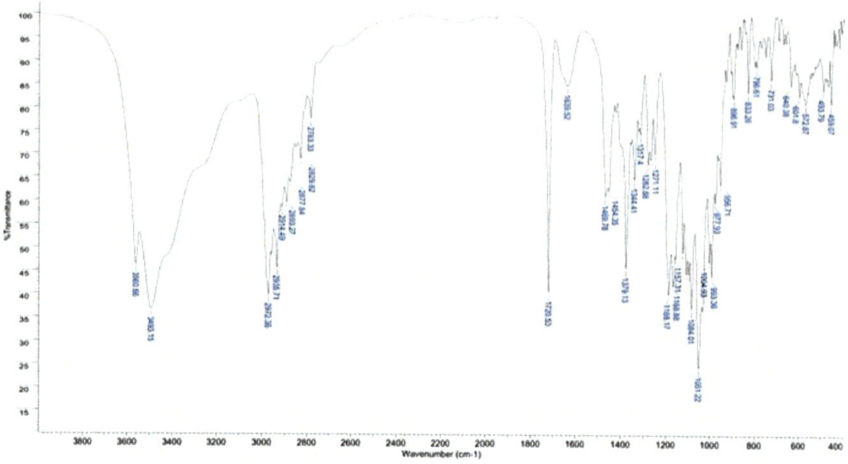

Fig. 3 Characteristic IR spectrum of an azithromycin sample

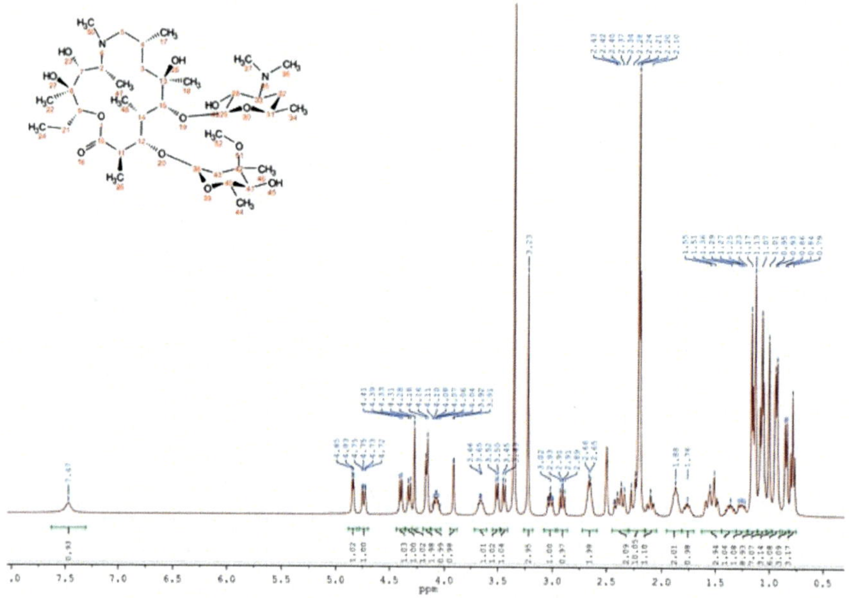

Fig. 4 1H NMR spectrum of an azithromycin sample in DMSO-d6

Mass Spectrometry

The monoisotopic mass of azithromycin samples was measured by coupled liquid chromatography-mass spectrometry (LC–MS/MS)[2] [8].

Reagents: trifluoroacetic acid ("for HPLC, ≥ 99.0%", Sigma-Aldrich, CAS No. 302031 or equivalent); acetonitrile (Gradient Grade for Liquid Chromatography, Merck CAS No. 1.00030 or equivalent).

Solution A. 0.50 ml of trifluoroacetic acid is added into a 1000 ml volumetric flask containing a small amount of water for chromatography. The flask is filled up to the mark with the water for chromatography, and the solution is stirred. The resulting solution is filtered through a membrane filter with a pore diameter of no more than 0.45 μm (if necessary) and degassed with ultrasound.

Solution B. Acetonitrile is filtered through a membrane filter with a pore diameter of no more than 0.45 μm (if necessary) and degassed with ultrasound.

Sample Solvent. Solutions A and B are mixed in a volume ratio of 50:50.

Test solution. Approximately 0.005 g (accurately weighed quantity) of the product is added to a 50 ml volumetric flask and dissolved in the sample solvent (an ultrasonic bath without heating can be used). The flask is filled up to the mark with the sample solvent and filtered through a membrane PTFE with a pore diameter of no more than 0.45 μm.

The analysis was carried out under the following conditions:

Chromatographic Conditions

- high-pressure liquid chromatograph;
- metal column filled with octadecylsilyl sorbent, 1.8 μm, 2.1 × 150 mm, for example, Zorbax SB-C18 (Agilent, USA), CAS No. 859700–902;
- column temperature: 35 °C;
- mobile phase flow rate: 0.35 ml/min;
- gradient elution regime (Table 1);
- volume of a sample: 5 μl.

Detection Conditions

- MS detector with heated electrospray and quadrupole mass analyzer;
- capillar temperature—350 °C;
- scanning mode—full ionic current;
- scanning range—35–500 m/z;

[2] RF, GPNR.1.2.1.1.0008.15 "Mass spectrometry", GPNR.1.2.1.2.0005.15 "High-performance liquid chromatography".

Table 1 Elution regime of azithromycin samples

Stage	Time, min,	Solution A,%	Solution B,%	Elution Regime
1	0	95	5	
2	1	95	5	isocratic
3	18	5	95	linear gradient
4	20	5	95	isocratic
5	20.1	95	5	linear gradient
6	25	95	5	isocratic, column equilibrium

- scanning frequency—0.5 s;
- scanning polarity—positive.

The sample solvent and the test solution are chromatographed consecutively.

LC–MS spectra were recorded using a liquid chromatograph with a Thermo Scientific TSQ Quantum Access MAX ™ mass selective detector (HESI ionization type) for positively charged ions.

Results and Discussion

The frequencies of the main absorption bands in the IR spectrum of the azithromycin certified RM of Series 1 are shown in Table 2.

IR spectrum (KBr), cm^{-1}: 459, 573, 731, 797, 833, 897, 957, 993, 1051, 1084, 1096, 1107, 1125, 1157, 1169, 1188, 1271, 1283, 1317, 1344, 1379, 1454, 1470, 1640, 1721, 2783, 2830, 2878, 2893, 2914, 2936, 2961, 2972, 3493, 3561.

Several broad bands in the range of 3273–3561 cm^{-1}, characteristic of the stretching vibrations of O–H, and bands in the range of 1084–1188 cm^{-1} corresponding to the stretching vibrations of the C–C–O bonds confirm the presence of alcohol groups in the compound. The latter lay in the same range with the bands assigned to stretching vibrations of C–N bonds of tertiary amino groups (993–1188 cm^{-1}), C–O–C bonds of ethers (1084–1125 cm^{-1}), and lactone bonds: C–C (= O) -O (about 1169 cm^{-1}) and C–C–O (about 1096 cm^{-1}). The absorption band of the C = O bond corresponding to the ester group is observed at about 1721 cm^{-1}. Asymmetric and symmetric stretching vibrations of C–H bonds assigned to methyl groups exhibit bands at 2972 and 2961 cm^{-1}, respectively, and bending vibrations—at 1454 (asymmetric) and 1379 cm^{-1} (symmetric). Stretching vibrations of C-H bonds of methylene and methine groups are observed as a group of bands in the range of 2783–2914 cm^{-1}. Bending vibrations of these bonds give several bands in the range of 1157–1344 cm^{-1} (wagging, twisting) and a band at 1470 cm^{-1} (scissoring).

Therefore, the position of main absorption bands observed in the IR spectra for different series of samples agrees with the chemical structure of azithromycin.

Table 2 IR absorption bands of a sample of substance 80,220

Absorption maximum, cm^{-1}	Vibration mode	Assignment
3273–3561	ν OH	-OH
2972	ν_{as} C – H	$-CH_3$
2961	ν_s C – H	$-CH_3$
2783–2914	ν C – H	$-CH_2$-, -C-H <
1721	ν C – H	> C = O
1470	δ_s-C – H (scissoring)	$-CH_2-$
1454	δ_{as}-C – H	$-CH_3$
1379	δ_s-C – H	$-CH_3$
1157–1344	δ_{as}-C – H (wagging, twisting)	$-CH_2-$
993–1188	ν C – C – O, ν C – N, ν C – O – C, ν C– (C = O) –O	– C – C – OH, –C – N <, – C – O – C– (simple ether), – C– (C = O) –O–, – C – C – O– (lactone)

*ν – stretching vibrations (ν_s - symmetric, ν_{as} - asymmetric); δ – bending vibrations (δ_s - symmetric, δ_{as} - asymmetric)

Table 3 presents the description and assignment of signals in ^{1}H NMR spectra. ^{1}H NMR spectrum (400 MHz, DMSO-d6) δ ppm: 0.79 (t, J = 7.4 Hz, 3H); 0.85 (d, J = 6.7 Hz, 3H); 0.90–0.97 (m, 6H); 1.01 (s, 3H); 1.03–1.11 (m, 7H); 1.11–1.20 (m, 9H); 1.26 (dd, J = 14.6; 8.0 Hz, 1H); 1.31–1.44 (m, 1H); 1.44–1.63 (m, 3H); 1.70–1.82 (m, 1H); 1.83–1.95 (m, 2H); 2.03–2.14 (m, 1H); 2.16–2.29 (m, 10H); 2.31–2.44 (m, 2H); 2.61–272 (m, 2H); 2.91 (dd, J = 9.4; 7.4 Hz, 1H); 2.98–3.07 (m, 1H); 3.23 (s, 3H); 3.44 (d, J = 8.0 Hz, 1H); 3.51 (d, J = 7.0 Hz, 1H); 3.60–3.71 (m, 1H); 3.91 (d, J = 2.0 Hz, 1H); 4.03–4.12 (m, 1H); 4.13–4.19 (m, 2H); 4.28 (s, 1H); 4.32 (d, J = 7.9 Hz, 1H); 4.40 (d, J = 7.3 Hz, 1H); 4.74 (dd, J = 10.0; 2.8 Hz, 1H); 4.84 (d, J = 4.7 Hz, 1H); 7.47 (bs, 1H).

A characteristic triplet with a spin–spin coupling constant J = 7.4 Hz at 0.79 ppm is assigned to the methyl group at C24, related to the substituent ethyl. The farthest downfield signal of methyl groups belongs to protons of methoxy group - OCH_3 (singlet at 3.23 ppm). Furthermore, a singlet at 1.01 ppm and a doublet with a spin–spin coupling constant J = 6.7 Hz at 0.85 ppm correspond to methyl groups. The signals of the other ten methyl groups overlap each other; the signals of the other protons are present in the spectrum as part of the various multiplets. Singlets at 4.28 and 7.47 ppm and doublets at 3.91 and 4.32 ppm on the spectrum are assigned to protons in the hydroxyl groups. The proton signal of another hydroxyl group is evidently part of one of the complex multiplets. The signals of two methine protons

Table 3 Description and assignment of signals in ^{1}H NMR spectrum of an azithromycin sample

Chemical shift, ppm	Multiplicity	SSCC J, Hz	Number of protons	Assignment
0.79	T	7.4	3	24
0.85	D	6.7	3	17
0.90–0.97	M	-	6	47, 48
1.01	S	-	3	22
1.03–1.11	M	-	7	26, 34, 32
1.11–1.20	M	-	9	18, 44, 46
1.26	DD	14.6, 8,0	1	3
1.31–1.44	M	–	1	21
1.44–1.63	M	–	3	3', 32', 43
1.70–1.82	M	–	1	21'
1.83–1.95	M	–	2	4, 14
2.03–2.14	M	–	1	5
2.16–2.29	M	–	10	36, 37, 50, 43'
2.31–2.44	M	–	2	5', 33
2.61–2.72	M	–	2	2, 11
2.91	DD	9.4, 7.4	1	41
2.98–3.07	M	–	1	28
3.23	S	–	3	52
3.44	D	8.0	1	7
3.51	D	7.0	1	15
3.60–3.71	M	–	1	31
3.91	D	2.0	1	49
4.03–4.12	M	–	1	40
4.13–4.19	M	–	2	12, 45
4.28	S	-	1	27
4.32	D	7.9	1	23
4.40	D	7.3	1	29
4.74	DD	10.0, 2.8	1	9
4.84	D	4.7	1	38
7.47	B S	–	1	25

*s - singlet, d - doublet, t - triplet, m - multiplet, dd - doublet of doublets, bs - broadened singlet

(at C29 and C38) are observed as two doublets (4.40 and 4.84 ppm). The signals assigned to these protons are in the upfield region of the spectrum, since all of them have an oxygen atom as a substituent in the β-position. In the same part of the spectrum, a doublet of doublets is observed corresponding to a proton at the carbon atom near the ester group (C9). A more detailed assignment of signals was performed

Table 4 Relative chemical shifts and signal assignment of ^{13}C NMR spectrum of an azithromycin sample

δ13C, ppm	Assignment	δ13C, ppm	Assignment
6.8	47, 48	44.8	11
9.1		48.8	52
11.0	22, 24, 26, 44	61.5	2, 31, 33, 40
14.8		64.8	
17.7		67.0	
18.5		68.8	5
20.99	21	70.6	28
21.03	17, 34, 46	72.6	8, 13, 42
21.5		72.8	
22.1		73.6	
26.1	4, 18	74.9	7, 9, 12, 41
27.4		76.4	
30.0	32, 43	77.4	
34.7		82.7	15
35.8	50	94.5	38
40.4	36, 37	102.2	29
41.7	14	177.1	10
41.9	3	–	–

using literature data. The overall integrated intensity, position, and multiplicity of signals agree with the substance structure, as well as the literature data [17].

^{13}C NMR spectrum (100 MHz, DMSO-d6) δ ppm: 6.8; 9.1; 11.0; 14.8; 17.7; 18.5; 20.99; 21.03; 21.5; 22.1; 26.1; 27.4; 30.0; 34.7; 35.8; 40.4; 41.7; 41.9; 44.8; 48.8; 61.5; 64.8; 67.0; 68.8; 70.6; 72.6; 72.8; 73.6; 74.9; 76.4; 77.4; 82.7; 94.5; 102.2; 177.1. Table 4 presents the assignment of signals in ^{13}C NMR spectra.

In the ^{13}C NMR spectrum, the number of signals is lower than that of magnetically nonequivalent carbon atoms in the molecule (Fig. 1). This can be explained by the overlapping of some signals, as observed from their intensity (presumably signals at 64.8 and 77.4 ppm). Moreover, the peak at 21.0 ppm is a superposition of two signals, one of them is related to the methylene group (C21) and was revealed using a DEPT135 experiment (Fig. 5).

Additionally, this spectrum allowed other methylene groups (C3, C5, C32, C43) and quaternary carbon atoms (C8, C10, C13, C42) to be identified and assigned. The farthest upfield of them (177.1 ppm) belongs to the carbonyl group carbon (C10). Signals at 94.5 and 102.2 ppm refer to carbon atoms bonded to two oxygen atoms (C28 and C29) (Fig. 6).

The position and multiplicity of signals in the ^{1}H and ^{13}C NMR spectra of an azithromycin sample are consistent with its chemical structure. The obtained ^{1}H and ^{13}C NMR chemical shifts corresponding to different series of azithromycin samples are similar (the discrepancy is lower than 0.01 ppm).

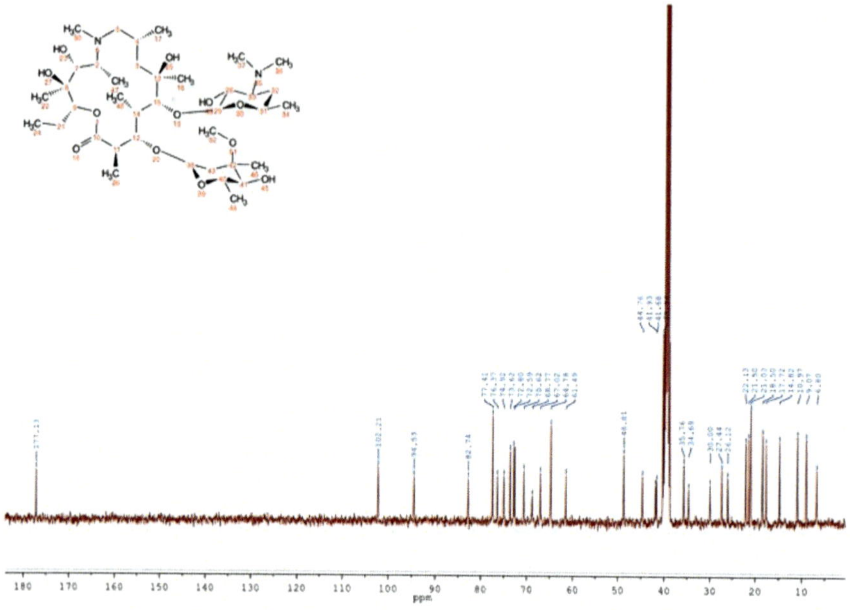

Fig. 5 ^{13}C NMR spectrum of an azithromycin sample in DMSO-d6

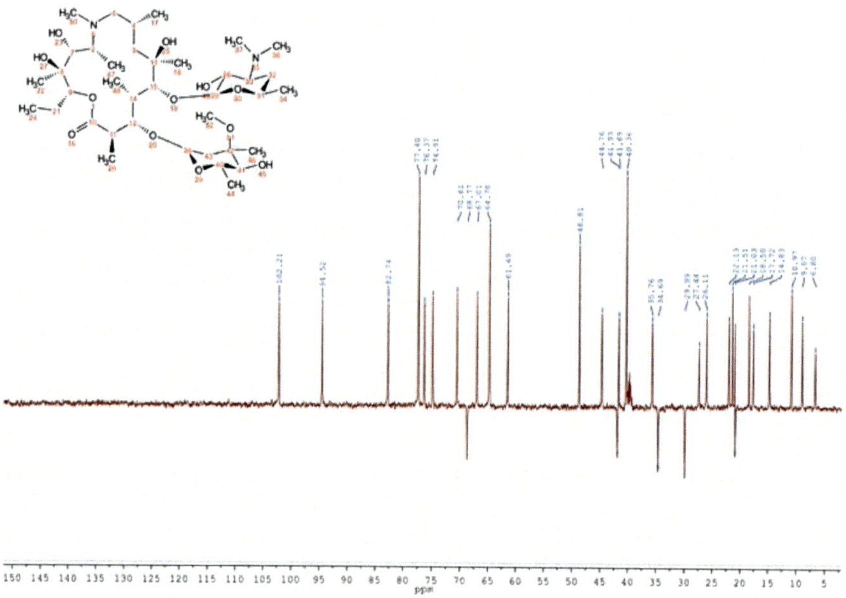

Fig. 6 ^{13}C DEPT135 NMR spectrum of an azithromycin sample in DMSO-d6

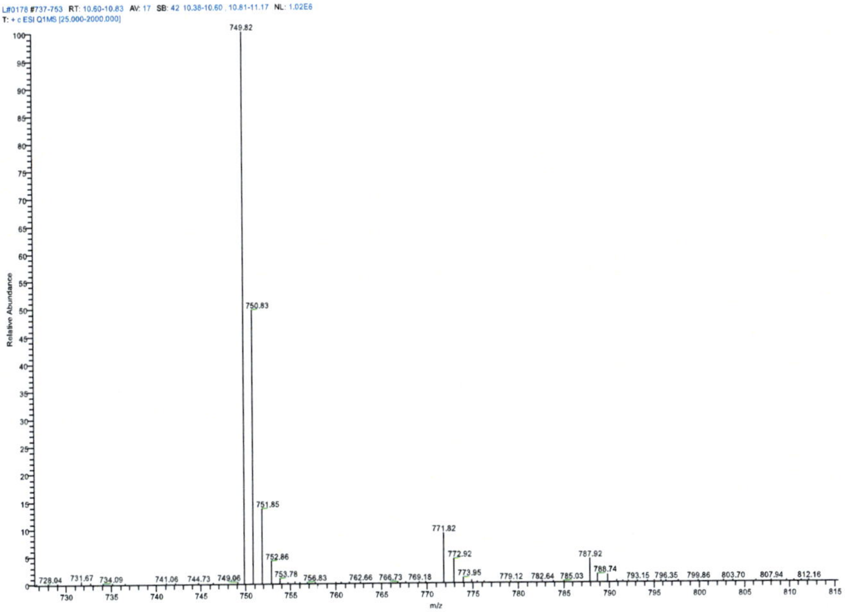

Fig. 7 Mass spectrum of azithromycin sample (HPLC–MS, HESI+)

A peak of protonated molecular ion m/z [C38H72N2O12 + H] + = 749.8 (Fig. 7) was detected on the mass spectrum of samples in all series and corresponds to calculated one [M + N] + = 749.5 [6].

Conclusion

A method for identifying azithromycin RMs using three different physicochemical methods was developed.

The structure of azithromycin was confirmed by the position of the main absorption bands in the IR spectrum. The position and multiplicity of signals in the ^{1}H and ^{13}C NMR spectra of azithromycin are consistent with its chemical structure. It was confirmed that ^{13}C NMR spectroscopy decoupled from protons is more convenient and informative for azithromycin identification. The mass spectrum of an azithromycin sample revealed a peak of protonated molecular ion m/z [C38H72N2O12 + H] + = 749.8, which also corresponds to the chemical structure of azithromycin.

The obtained results allowed the structure of azithromycin to be reliably identified by combining several physicochemical methods.

The standardization of the azithromycin composition remains a promising research topic, since the development of a RM, in addition to identification, includes the following: establishing certified values of the quantities describing the content of the components of interest in the substance, as well as and investigating homogeneity and stability.

Acknowledgements The authors express their gratitude to the First Deputy Director Sergey Viktorovich Polyakov, researchers of the Laboratory for Analytical Research Methods of the State Institute of Drugs and Good Practices, as well as employees of the National Center for Reference Materials, LLC.

Author Contributions V.I. Gegechkori and O. Yu. Schepochkina: discussion, analysis, and processing of experimental data; text preparation.

N.A. Shulga: analysis of literature data.

N.V. Gorokhovets: collection and analysis of experimental data, text preparation.

V.N. Shestakov and A.A. Levko: research supervision.

N.N. Chadova: text preparation.

Conflict of Interest The article was prepared on the basis of a report presented at the IV International Scientific Conference "Reference Materials in Measurement and Technology" (St. Petersburg, December 1–3, 2020). The article was admitted for publication after the abstract was revised, the article was formalized and the review procedure was carried out.

The version in the Russian language is published in the journal "Measurement Standards. Reference Materials" 2021;17(3):21–34. https://doi.org/10.20915/2687-0886-2021-17-3-21-34.

References

1. Shchepochkina OY, Gegechkori VI, Prokof'eva VI, Chepilo DA, Levko AA, Chadova NN, Shestakov VN (2020) Modern approaches to the development of standard samples of drugs. Pharm Chem J 54(7):761-765. https://doi.org/10.1007/s11094-020-02267-y
2. Dorofeev VL, Arzamastsev AP (2010) Reference samples for pharmacopoeial analysis. Probl Biol Med Pharm Chem 8(5):6–10 (In Russ)
3. GOST R ISO 9001-2015 (2020) Quality management systems. Requirements. Standartinform, Moscow (In Russ)
4. GOST ISO Guide 35-2015 (2016) Reference materials. General and statistical principles for certification. Standartinform, Moscow (In Russ)
5. Veselov AV, Kozlov RS (2006) Azithromycin: modern aspects of clinical application. Clin Microbiol Antimicrob Chemother 8(1):18–32 (In Russ)
6. Exact Mass Calculator, Single Isotope Version. https://www.sisweb.com/referenc/tools/exactmass.htm
7. State Pharmacopoeia of the Union of Soviet Socialist Republics, 10th ed. Medicine, Moscow (1968), 1079p (In Russ)
8. State Pharmacopoeia of the Russian Federation, XIV ed. Moscow. http://www.femb.ru
9. European Pharmacopoeia (2017) In: 8th Ed European department for the quality of medicines. European Department for the Quality of Medicines, Strasbourg
10. British Pharmacopoeia 2016. H. M. Stationary Office, London, 2015, 1276p
11. The United States Pharmacopeia 40-National Formulary 35, Reissue (2018)
12. Gegechkori VI, Shchepochkina O, Pyatin BM, Grushevskaya LN (2015) Development of a method for the determination of heavy metals in standard samples of preparations of peptide structure using the method of atomic absorption spectrometry. Nat Techn Sci 10(88):201–205 (In Russ)

13. Gegechkori VI, Kokorekin VA, Shchepochkina OY, Piatin BM, Grushevskaya LN, Avdunina NI, Gaevaya LI (2016) Application of ^{1}H and ^{13}C nmr spectroscopy in the analysis of reference standards of dilept and GB-115, medicinal products of peptide structure. Russ J Biopharm 8(3):37–43 (In Russ)
14. Gegechkori VI, Shchepochkina OY, Rodionova GM, Grushevskaya LN, Piatin BM (2015) Use of gas chromatography in quality control of reference standards of medicinal products of peptide structure. Russ J Biopharm 7(6):43–48 (In Russ)
15. GOST ISO Guide 33-2019 (2019) Reference materials. Good practice in using reference materials. Standartinform, Moscow (In Russ)
16. ISO/TR 10989:2009: Reference materials–guidance on, and keywords used for, RM categorization. https://www.iso.org/standard/46526.html
17. Brennan RJ, Barber J (1992) Full assignments of the ^{13}C and ^{1}H NMR spectra of azithromycin in buffered D_2O and DMSO-d_6. Magn Reson Chem 30(4):327–333. https://doi.org/10.1002/mrc.1260300410

Vladimir I. Gegechkori Cand. Sci. (Farm.), Associate Professor of the Department of Pharmaceutical and Toxicological Chemistry named after A.P. Arzamastsev Institute of Pharmacy A.P. Nelyubina, I.M. Sechenov First Moscow State Medical University of the Ministry of Health of the Russian Federation (Sechenov University), State Institute of Drugs and Good Practices.

Nikolai A. Shulga Laboratory Assistant at the Department of Pharmaceutical and Toxicological Chemistry named after A.P. Arzamastsev Institute of Pharmacy A.P. Nelyubina, I.M. Sechenov First Moscow State Medical University of the Ministry of Health of the Russian Federation (Sechenov University), State Institute of Drugs and Good Practices.

Olga Yu. Shchepochkina Cand. Sci. (Farm.), Associate Professor of the Department of Pharmaceutical and Toxicological Chemistry named after A.P. Arzamastsev Institute of Pharmacy A.P. Nelyubina, I.M. Sechenov First Moscow State Medical University of the Ministry of Health of the Russian Federation (Sechenov University).

Neonila V. Gorokhovets Cand. Sci. (Biol.), Leading Researcher of the Laboratory of Molecular Biology and Biochemistry, I.M. Sechenov First Moscow State Medical University of the Ministry of Health of the Russian Federation (Sechenov University).

Anatolii A. Levko Cand. Sci. (Med.), Deputy Head of the Department for Inspection of Production of Medicines and Expertise, State Institute of Drugs and Good Practices.

Nataliia N. Chadova Cand. Sci. (Chem.), Head of the Department for Inspection of Production of Medicines and Expertise, State Institute of Drugs and Good Practices.

Vladislav N. Shestakov Director, State Institute of Drugs and Good Practices.

Application of Certified Reference Materials of Absorbed Dose for Process Validation of Irradiation of Medical Supplies and Food Products

Alexander A. Gromov, Alexander P. Zhanzhora, and Oleg I. Kovalenko

Abstract At present, radiation technologies are most commonly applied for sterilization of medical supplies and irradiation of food products. The results of radiation treatment cannot be fully verified by subsequent non-destructive testing. Therefore, irradiation processes and equipment used for radiation treatment should be subject to mandatory validation and certification. In this paper, we present the results obtained during process validation of radiation-technological installations based on electron accelerators by investigating absorbed dose in irradiated products.

Keywords Reference materials · Absorbed dose · Radiation treatment · Ionizing radiation · Food products · Medical devices

Introduction

According to the Food Security Doctrine of the Russian Federation,[1] food security is one of the main directions in ensuring the national security of the country. The strategic goal of food security is to provide the country's population with safe agricultural and food products.

---

[1] On the approval of the Food Security Doctrine of the Russian Federation: Decree of the President of the Russian Federation of January 30, 2010 No.120. Official Internet portal of legal information. Available at: http://pravo.gov.ru/proxy/ips/?docbody=&firstDoc=1&lastDoc=1&nd=102135612.

A. A. Gromov · A. P. Zhanzhora (✉) · O. I. Kovalenko
All-Russian Scientific Research Institute of Physical, Technical and Radio Engineering Measurements (VNIIFTRI), Mendeleevo, Moscow Region, Russia

A. A. Gromov
e-mail: gromov_a@vniiftri.ru

O. I. Kovalenko
e-mail: koi@vniifnri.ru

S. V. Medvedevskikh et al. (eds.), *Reference Materials in Measurement and Technology*,
https://doi.org/10.1007/978-3-031-06285-8_5

The report of the International Food and Agriculture Organization of the United Nations (FAO), published in 2014[2] claims that, on average, 30% of all products manufactured globally (about 1.3 billion tons) are subject to spoilage. This amount includes 40–50% of fruit, vegetable, and root crops, 20% of meat and 35% of fish. Almost a third of all food products is lost every year due to spoilage. Millions of people suffer from food and waterborne gastrointestinal infections [1–3].

These problems are relevant for all countries, including Russia. The Russian Association for Electronic Communications (RAEC) and the consulting company TIARCENTER have calculated the amount of food discarded annually. The numbers turned out to be impressive: shops, horticultural centers, households, etc., annually destroy 17 million tons of food worth 1.6 trillion rubles. If this food is distributed among all Russian people, each will receive about 120 kg of food per year [4].

Sterilized single-use medical products are widely used not only in the medical practice of hospitals, but also by a wider population. Today, industrially-produced medical supplies are largely sterilized by the methods of heat, gas, and radiation treatment. Among other methods, radiation sterilization is considered to be the most effective, safe, environmentally friendly, and reliable treatment [5–7].

An assortment of medical supplies subject to radiation sterilization includes hundreds of types: injection syringes, blood service products, various types of catheters, needles, surgical sutures, surgical drapes, dressings, alloplants, plasma filters, blood collection tubes, test containers, and many others. Although more than 300 facilities produce such products in Russia, a large number of medical products have to be purchased abroad.

Irradiation technologies used for treating medical and food products have many similar features. Thus, in both cases, the aim is to inactivate the microflora to safe levels. To this end, radiation-technological installations (RTI) based on radionuclide sources and electron accelerators are used. The most important parameter determining the efficiency of such processes is the absorbed dose (AD) of ionizing radiation. The accuracy, stability, reliability, and reproducibility of this parameter defines the quality and safety of irradiated products.

Importantly, radiation sterilization and irradiation of food products are specific processes, since their effectiveness cannot be fully verified (confirmed) by subsequent non-destructive testing. Therefore, all such treatment procedures require validation and certification.

In this article, we investigate the use of ionizing radiation for process validation of destructing all forms of viable microorganisms using certified reference materials (CRMs) of absorbed dose traceable to the State Primary Special Standard of the Unit of Absorbed Dose of Intense Photonic, Electronic and Beta—Radiation for Radiation Technologies GET 209-2014.

[2] The State of Food Insecurity in the World 2014. Food and Agriculture Organization of the United Nations FAO. Available at: http://www.fao.org/publications/sofi/2014/ru/.

Materials and Methods

In Russia, products are irradiated using specialized radiation-technological installations (RTI), mainly using an electron accelerator as a radiation source of the following types: UELR-10-15S, UELR-10-10-C2, UELV-10-10-S-70, UELV-10-10-T, etc. The entire process is safe and reliable, provided that all technological requirements are met. Since the results of irradiation treatment cannot be fully verified by subsequent non-destructive testing of the finished products, all radiation-technological processes and applied equipment should undergo validation and certification.

GOST R 56,431-2015/GHNF/SG/N99-10:2004 [8] provides manufacturers with general guidelines concerning validation of technological processes. According to the definition given in [8], process validation is "... providing objective evidence that the process consistently produces a result or product that meets the established requirements."

As a rule, manufacturing facilities apply commercial RTIs to treat their products by ionizing radiation. Sterilization of medical supplies and irradiation of food products are similar processes, implying the inactivation of microorganisms and fungi. Therefore, the approaches used for validating these processes are largely similar.

The process validation of radiation sterilization of medical products is detailed in [9]. When validating these processes, much attention should be paid to both the installed equipment (operational qualification) and RTIs (maintenance qualification) in terms of verifying absorbed dose in treated (irradiated) objects.

Research Object

The validation of equipment implies establishing: the energy of electrons, which should not exceed 10 meV; the non-uniformity of the distribution of the current density of the electron beam over the scan width of the accelerator beam, which should not exceed 10%; the stability of the accelerator and the RTI as a whole under various operating conditions.

The validation of an RTI in terms of absorbed dose implies determining: the possibility of irradiating objects in the established range of absorbed doses (with the minimum and maximum values of the absorbed dose in the product); the range of absorbed doses at a checkpoint; proportionality coefficients k_{min} and k_{max}, determining the relationship between the dose at a checkpoint and the minimum and maximum values of absorbed dose in the product under regulated irradiation conditions; minimum V_{min} and maximum V_{max} speed of moving products under the electron beam.

Equipment

In the process of research, the following measuring instruments and equipment were used:

- the State Primary Special Standard of the Unit of Absorbed Dose of Intense Photonic, Electronic and Beta -Radiation for Radiation Technologies GET 209-2014 [10];
- the certified reference material (hereinafter referred to as the CRM) of the absorbed dose of photon and electron ionizing radiation (copolymer with phenazine dye) CRM AD (F) E-5/50 registered by the Federal Information Fund for Ensuring the Uniformity of Measurements as GSO 7904-2001 [11]. The permissible values of the relative error of the CRM value comprise 3% with a confidence level of 0.95 for the range of absorbed doses in water (hereinafter referred to as AD) from 5 to 50 kGy. The error in measuring the absorbed dose using the CRM does not exceed 7% (P = 0.95), provided that the operation requirements are observed.
- a spectrophotometer Specord M40 for measuring optical density in the visible and ultraviolet spectral range ($\delta \leq 0.1\%$, P = 0.95);
- a wedge device designed for measuring the energy of accelerated electrons;
- a device for measuring the optical density of extended CRMs with a step from 0.1 to 20 mm.

Results and Discussion

The validation of an accelerator is an important stage in the validation of irradiation processes, since its accurate operation determines the quality and safety of finished medical and food products. Validation was carried out according to the methods described in [12, 13].

For irradiation of medical and food products, electron radiation with a maximum energy of no more than 10 meV is used, since this regime produces no induced radioactivity in the irradiated object. Therefore, it is of crucial importance to ensure that the electron energy at a given RTU does not exceed the specified value.

The determination of the average E_{av} and the most probable E_p of the electron energy in the irradiation zone was carried out according to the empirical dependence of the energy on the electron free path in aluminum:

$$E_p = 0.2 + 5.09R_0(Al)\, at\ 1\, MeV \leq E_{av} \leq 50\, MeV, \tag{1}$$

$$E_{\text{av}} = 6.2R_{50}(Al) \text{ at 1 MeV} \leq E_{av} \leq 12 \text{ MeV}, \tag{2}$$

where R_0 and R_{50} are the electron free path in aluminum, cm (see Fig. 1).

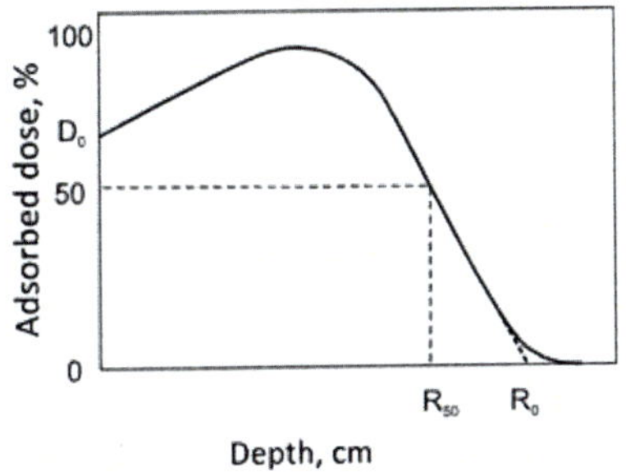

Fig. 1 Typical distribution of the absorbed dose across the depth of the absorber

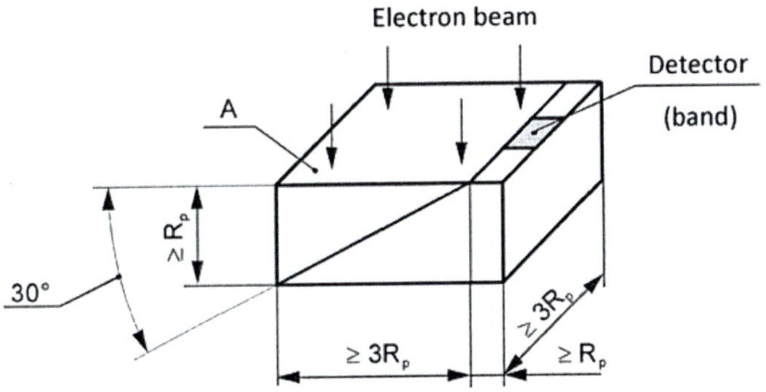

Fig. 2 A wedge device for measuring the energy of electrons, where R_a is the actual free path of electrons, cm

The free path R_0 and R_{50} were determined from the AD distribution in terms of absorber depth. For measurements, we constructed a wedge device—a thick aluminum plate cut at an angle of 30°, the dimensions of which are shown in Fig. 2.

The wedge device was manufactured from pure aluminum grade A7 GOST 11,069–2001. A groove was made in the wedge device (Fig. 2), in which the CRM in the form of a tape was inserted to the entire depth of the plate. To reduce the random measurement error, at least three CRM tapes were placed side by side at the same time.

For irradiation, the device was placed on the conveyor in the middle of the beam scan, since the front surface of the wedge (plane *A* in Fig. 2) should be perpendicular to the electron beam axis and be placed at the same distance from the exit window of the accelerator as the front surface (with respect to the beam) of a box with the objects under irradiation.

The irradiation was carried out by passing the device with the CRM through the irradiation zone once. We measured AD obtained by the CRM with the required step (it was 0.014 mm in our experiment) along the entire length of the tape followed by plotting its distribution across the absorber depth. In order to ignore the CRM thickness when determining the distance from the input surface to the CRM, the

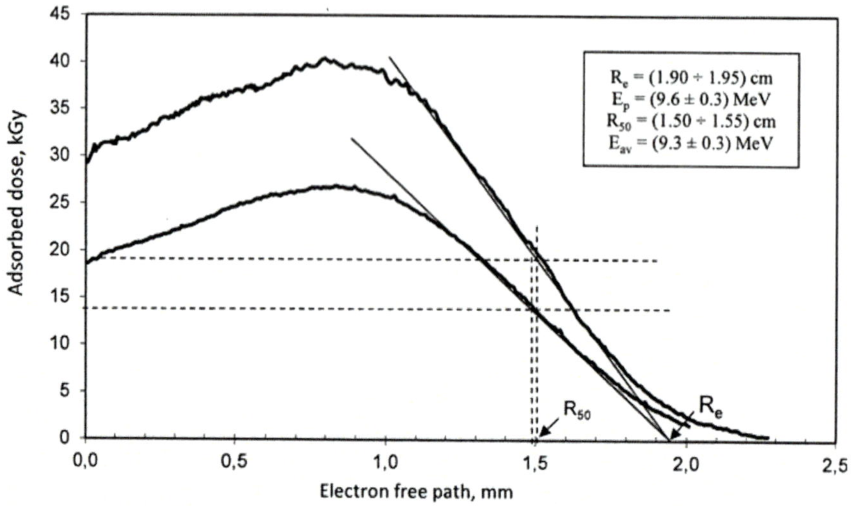

Fig. 3 Depth distribution of the absorbed dose in aluminum. The operating mode of the RTI during validation: the specified electron energy $E_0 = 9.4$ meV, the average electron beam current $I_n = 730$ μA, the conveyor line speed V = 0.60 and 0.90 m/min, the scan width SW = 40 cm, one-sided irradiation

CRM must be located as a recording layer with a thickness of 0.015 ± 0.002 mm, facing the incident electron beam. Therefore, under an electron free path of 1.5–2 cm, the layer thickness can be neglected.

The plots of the absorbed dose distribution in aluminum were used to determine the actual free path of electrons R_a and the half-attenuation layer R_{50}. Then the E_p and E_{av} of electron energies were calculated using formulas (1) and (2). As an example, Fig. 3 shows the AD distribution across the depth of an aluminum absorber.

If the maximum electron energy does not exceed 10 meV, the irradiation of objects can be considered safe.

Another important controllable parameter of an RTI is the uniformity of movement of the irradiated objects (stability of the speed of moving the irradiated objects, V) along the irradiated zone. To evaluate this parameter at different values of the conveyor speed, the distribution of the absorbed dose D (V) was measured on a flat surface under the most unfavorable conditions, for example, on the box surface closest to the exit window of the accelerator, along the direction of table motion during one pass of the irradiation zone. Irradiation was carried out by passing the CRM through the irradiation zone once. The AD was measured across the entire length of the tape with a step of 5 mm. The irradiation scheme is shown in Fig. 4.

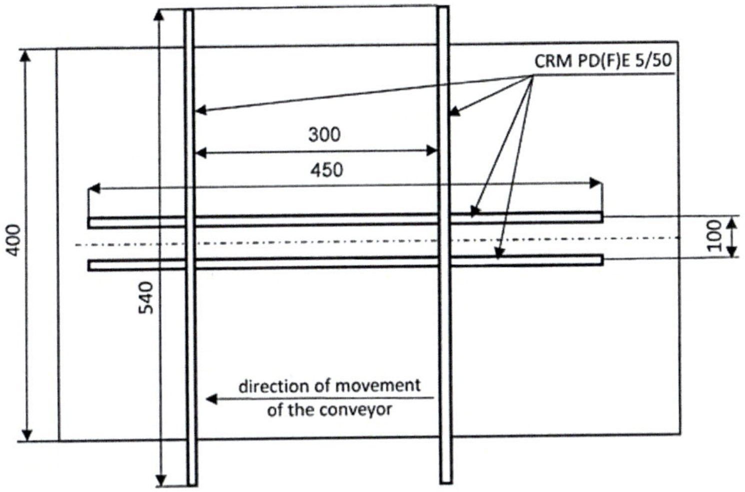

Fig. 4 Scheme of CRM irradiation when evaluating the uniformity of movement of the irradiated objects and the absorbed dose distribution over the scan width

The measurements were carried out under several operating modes of the RTI, aiming to cover the entire operating range in terms of electron energies (E_0) and the current of the deflecting magnet of the beam scan (scan width) (I_{mag}). An example of measurements is shown in Fig. 5. Table 1 presents the measurement results: the average AD values in the uniform section $\overline{D}$, the mean-square deviations of the observation results *S(D)* and the values of the coefficient $K_0 = \overline{D} \cdot V_k / I_n$ and $S(K_0)$.

S(D) characterizes the stability of the conveyor speed, while $S(K_0)$ determines the linearity and reproducibility of the systems controlling the conveyor speed, beam scan width, and electron beam current of the RTI accelerator. These characteristics should not exceed 5%, which was the case for the installation under study.

The system of beam scanning should ensure a uniform irradiation of an object in the transverse direction of the object movement. According to [14, 15], the irregularity of irradiation across the width should not exceed 10%. Measurements of the scan width (the length of a uniform section of the PD distribution on the surface) were carried out simultaneously with the previous measurements (Fig. 4). The measurements are exemplified in Table 1 and Fig. 8. The length of the uniform section was 370 mm. Under a transport box width of 370 mm, the electron beam covers the surface of the box completely. The coefficient of unevenness does not exceed 1.1, which agrees well with [9, 10].

Provided that the measurement results are positive, a Certificate of Validation is issued.

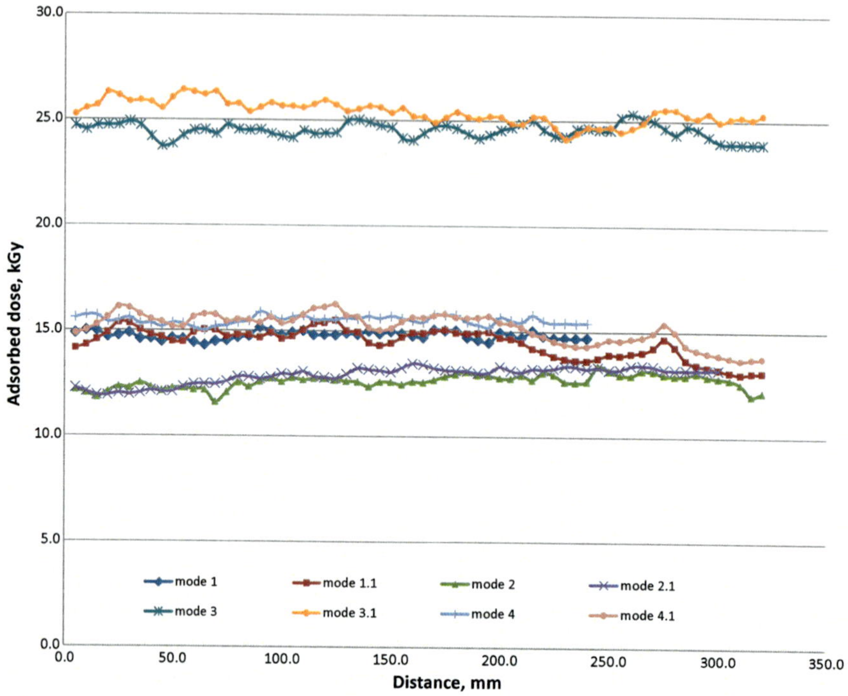

Fig. 5 Measurement of the absorbed dose when assessing the uniformity of movement of irradiated objects under different irradiation regimes

Table 1 Modes of RTI operation and the results of measuring: mean value of absorbed dose $\overline{D}$, mean-square deviation of the observation results *S(D)*, and coefficient K_0

Irradiation Mode	1	2	3	4
Pulse frequency, F_p, kHz	0.201	0.201	0.100	0.169
Average beam current, I_b, A	0.955	0.955	0.477	0.798
Speed, V_k, m/min	1.25	0.65	0.65	0.9
Average value of the absorbed dose, $\overline{D}$, kGy	12.6 and 12.9	24.5 and 25.4	12.3 and 12.5	14.6 and 14.8
S(D), %	3	2	4	3
$K_0 = \overline{D} \cdot V_k / I_b$	16.5 and 16.9	17.3 and 16.5	16.9 and 16.5	17.0 and 16.7
K_0 and $S(K_0)$	$K_0 = 16.8$ $S(K_0) = 1.7\%$			

Validation of RTIs based on electron accelerators according to the value of absorbed dose in irradiated products is carried out according to [7] (for medical products) and [8] (for food products). Each type of products is assigned a range of absorbed doses: minimum D_{min} and maximum allowable D_{max}, across which the irradiated products will meet the declared requirements for safety and quality during the specified shelf life.

Dosimetry was performed as follows. The CRM tape was evenly distributed throughout the volume of a box filled with products and placed at a checkpoint. Thus, under the regime of two-sided irradiation, the CRM was placed at depths (H) of 0.0, 0.25, and 0.5 of the box height. The respective depths under the regime of one-sided irradiation were 0.0, 0.25, 0.5, 0.75 and 1. A checkpoint, as a rule, is a small area located on the outer surface of a transport box and designated for radiation monitoring of the entire process.

During irradiation of an object, the control levels are the boundaries of permissible values of the absorbed doses of ionizing radiation at a checkpoint. These boundaries, established during certification of an RTI in terms of absorbed dose for different product types, are used for radiation monitoring of the irradiation process.

Figure 6 shows the CRM tape located at the checkpoint on the surface of a transport box. Figure 7 demonstrate the placement of the CRM tape inside a box with products for irradiation (Fig. 8).

As an example, Fig. 9 presents the results of measuring the AD in the product "Medical diagnostic disposable sterile gloves", whose regulatory requirements include: sterilizing dose $D_{st} = 17.5$ kGy and maximum permissible dose $[D]_{max} = 50$ kGy.

Fig. 6 The CRM at the checkpoint

Fig. 7 CRM placement in a transport box with products

Products, each in an individual sealed packaging, were packed by 40 pieces in group packages —cardboard boxes measuring 280 × 190x155 mm. These packages were stacked by 8 pieces in cardboard boxes measuring 610 × 380x285 mm. The CRM was placed inside the box between the products, as well as at a checkpoint. The products were then subjected to two-sided irradiation. Table 2 shows the RTI operation mode during the process of validation.

The AD at the checkpoint, D_{check}, is associated with the minimum and maximum AD values as follows:

$$D_{min} = D_{check} * K_{min,} \tag{3}$$

$$D_{max} = D_{check} * K_{max} \tag{4}$$

where $K_{\min} = \overline{D}_{\min}/\overline{D}_{check}$ and $k_{\max} = \overline{D}_{\max}/\overline{D}_{check}$ are proportionality coefficients depending on irradiation conditions, object type, and location of the checkpoint. Table 3 presents the minimum D_{min} and maximum D_{max} AD values in products, AD at the checkpoint $D_{check,}$ and their ratios.

Based on the results obtained (Table 3), the RTI operation mode was established. For the modes specified in Table 2, we calculated a range of permissible values of the conveyor speed (see Table 4), in which the lower limit of the confidence interval

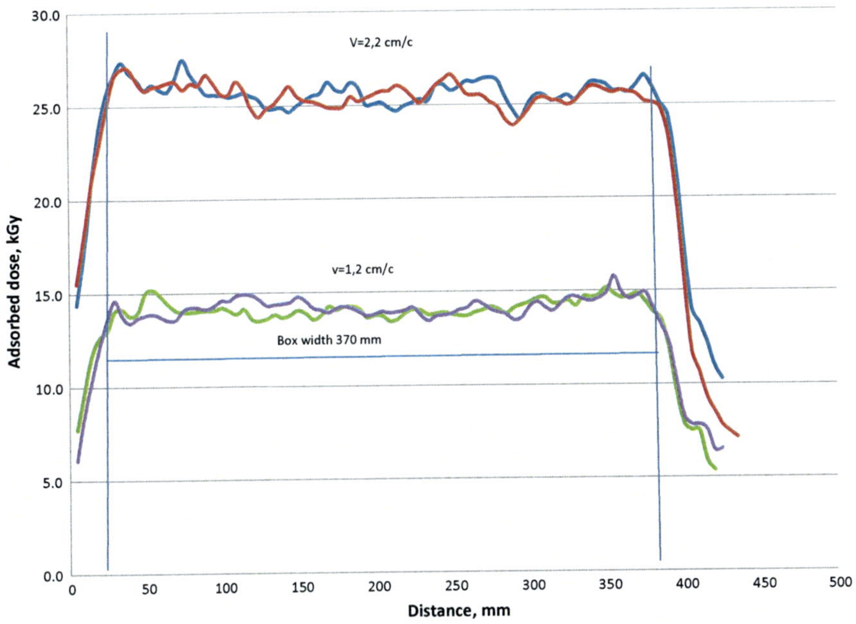

Fig. 8 Distribution of the absorbed dose on the box surface across the scan width. The mode of RTI operation during certification: $E_0 = 9.4$ meV, $I_b = 730$ μA, $V = (0.60 \div 2.0)$ m/min, width = 40 cm, one-sided irradiation

of D_{min} is not less than the sterilizing dose $D_{st} = 17.5$ kGy and the upper limit D_{max} is not higher than the maximum permissible dose $[D]_{max} = 50$ kGy.

Metrological studies of RTIs based on electron accelerators using the value of absorbed dose in irradiated products are recommended for confirming their suitability/unsuitability for radiation sterilization of medical supplies or irradiation of food products.

Conclusion

Using a practical example, we have demonstrated the applicability and convenience of certified reference materials of absorbed dose for verifying both the operation of radiation-technological installations based on electron accelerators and the efficiency of irradiation processes. The described approach can be used by manufacturers of medical supplies and food products for verifying radiation treatment processes.

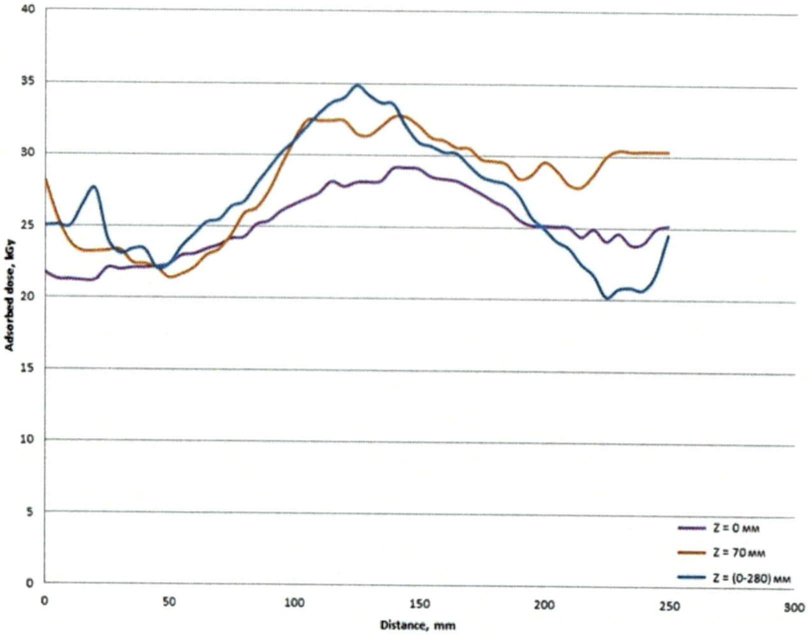

Fig. 9 Distribution of the absorbed dose in the product "Medical diagnostic disposable sterile gloves"

Table 2 Mode of RTI operation during validation of the product "Medical diagnostic disposable sterile gloves"

Conveyor speed, V_c, cm/s	1.8 ± 0.1
Beam current (monitor–electron collector), I_{mon}, mA	0.77 ± 0.02
Most probable electron energy, E_p, MeV	9.2
Current of deflecting sweep magnet, I_m, A	$-34.0 \div 30.0$

Table 3 Average absorbed dose values in the products under irradiation at the checkpoint D_{min}, D_{max}, D_{check}, as well as the proportionality coefficients K_{min} and K_{max}

Product	$\overline{D}_{\text{check}}$ kGy	$\overline{D}_{\text{min}}$ kGy	$\overline{D}_{\text{max}}$ kGy	K_{min}	K_{max}
Diagnostic gloves	25.2 ± 1.8	20.2 ± 1.4	41 ± 3	0.80 ± 0.05	1.6 ± 0.1

Table 4 Established ranges of conveyor speed and absorbed dose values in the products under irradiation and the checkpoint

Product	$[V_c]_{max} - [V_c]_{min}$, cm/s	Range of AD values in the products, D, kGy, at $V = [V]_{max}$	Range of AD values at the checkpoint, D_{check}, kGy, at $V = [V]_{max}$
Diagnostic gloves	2.0–1.9	17.5–48.4	22–30

Acknowledgements The authors express their gratitude to the reviewer of the article for valuable comments and suggestions.

Author Contributions A. A. Gromov: Research conceptualization, methodology design, experimental research supervision.

A. P. Zhanzhora: Methodology design, literature review, experimental research supervision, data analysis, text preparation.

O. I. Kovalenko: Literature review, experimental studies, data analysis.

Conflict of Interest The article was prepared on the basis of a report presented at the IV International Scientific Conference "Reference Materials in Measurement and Technology" (St. Petersburg, December 1–3, 2020). The article was admitted for publication after the abstract was revised, the article was formalized and the review procedure was carried out.

The version in the Russian language is published in the journal "Measurement Standards. Reference Materials" 2021;17(4).

References

1. Martynova A (2018) "Superpower" of ionizing radiation: making food safe. Atomic Expert Netw J 1(62). https://atomicexpert.com/foodradiology (In Russ)
2. Code Alimentarius: The irradiated food: Russ. ed., Izdatel'stvo «Ves' Mir», Moscow (2007), 21p (In Russ)
3. Majeed A, Muhammad Z, Majid A, Shah A, Hussain M (2014) Impact of low doses of gamma irradiation on shelf life and chemical quality of strawberry (Fragariaxananassa) cv.' Corona'. J Animal Plant Sci 24(5):1531–1536
4. Sigida A (2019) Delays can feed 30 million Russians. World NEWS Online Mag 43(1347). https://mirnov.ru/ekonomika/potrebitelskii-rynok/prosrochkoi-mozhno-prokormit-30-mln-rossijan.html (In Russ.)
5. Kalashnikov VV, Gordeev AV, Pavlov EP, Bushmanov AYu (2014) Development and application of radiation sterilization method in federal medical and biophysical centre N.A. A.I. Burnazyan. Saratov J Med Sci Res 10(4):844–849 (In Russ)
6. Fairand BP (2001) Radiation sterilization for health care products: X-ray, gamma, and electron beam. CRC Press, Boca Raton. https://doi.org/10.1201/9781482286205
7. Silindir M, Özer AY (2009) Sterilization methods and the comparison of E-beam sterilization with gamma radiation sterilization. FABAD J Pharm Sci 34:43–53
8. GOST R 56431-2015/GHNF/SG/N99-10:2004 Quality management system. Medical devices. Process validation guidance. Standardinform, Moscow, 2020 (In Russ.)
9. Gromov AA, Zhanzhora AP, Kovalenko OI (2020) The safety of single-use medical devices of the category "sterile." Qual Manage Med 4:109–113 (In Russ.)
10. Aleikin VV, Generalova VV, Gromov AA, Kovalenko OI (2019) The state primary special standard of the unit of absorbed dose rate of intense photon, electron and beta radiation for radiation technology GET 209–2014. Mesure Tech (7):10–13. https://doi.org/10.32446/0368-1025it.2019-7-10-13 (In Russ.)
11. Doynikov AS, Zhanzhora AP, Kovalenko OI, Tenishev VP, Hrapov FI, Yurov LV (2018) Analysis of the existing regulatory framework to ensure the uniformity of measurement for reference materials and development of proposals for its harmonization with international law. Al"manac Mod Metrol (13):145–188 (In Russ)
12. RMG 136-2016 (2016) State system for ensuring the uniformity of measurements. Radiation-technological facilities with electron accelerators for medical products sterilization. Certification techniques. Standardinform, Moscow (In Russ)

13. RMG 146-2019 (2019) State system for ensuring the uniformity of measurements. Radiation-technological facilities with electron accelerators for foodstuffs radiation processing. Methods of certification on absorbed dose in products. Standardinform, Moscow (In Russ)
14. GOST 8.664-2019 (2019) State system for ensuring the uniformity of measurements. Foodstuffs Foodstuffs radiation processing. Requirements to dosimetry provision. Standardinform, Moscow (In Russ)
15. GOST 8.651-2016 (2016) State system for ensuring the uniformity of measurements. Medical products. Radiation sterilization. Dosimetric techniques. Standardinform, Moscow (In Russ)

Alexander A. Gromov Acting Head of the laboratory, Senior Researcher-Scientific Keeper, All-Russian Scientific Research Institute of Physical, Technical and Radio Engineering Measurements (VNIIFTRI).

Alexander P. Zhanzhora Senior Researcher, All-Russian Scientific Research Institute of Physical, Technical and Radio Engineering Measurements (VNIIFTRI).

Oleg I. Kovalenko Cand. Sci. (Eng.), Acting Head of the Research Department of Metrology of Ionizing Radiation, All-Russian Scientific Research Institute of Physical, Technical and Radio Engineering Measurements (VNIIFTRI).

Development of Certified Reference Material Solutions for qNMR and Instrument Performance Qualification

Christine Hellriegel, Alexander Rueck, Kathrin Breitruck, and Markus Obkircher

Abstract Over the last decades quantitative NMR (qNMR) spectroscopy has become an important tool for the content determination of organic substances and the quantitative evaluation of impurities. Since the signal intensity is directly proportional to the number of protons contributing to the resonance, qNMR is considered as a relative primary method [1, 2, 3]. Quantitative NMR in combination with metrological weighing was optimized to show the power of the measurement method [4]. Traceability to the SI is achieved using primary Reference Materials from the National Institute of Standards and Technology (NIST) [5] and the National Metrology Institute of Japan (NMIJ) [6]. Neat material CRM for qNMR is already well established, and here we present the development of new CRM in deuterated solvent, which are ampuled and ready-to-use. Furthermore, a comparison of the results between qNMR measurements and gravimetric results of the preparation of the solutions under an ISO 17034 and ISO/IEC 17025 accreditation is presented.

Keywords qNMR · CRM (Certified reference material) · Metrological traceability · Accreditation · Certification · Instrument performance qualification · ISO 17034 · ISO/IEC 17025 · Gravimetry · Quantitative nuclear magnetic resonance · Measurement uncertainty

---

C. Hellriegel (✉) · A. Rueck · K. Breitruck · M. Obkircher
Sigma-Aldrich Production GmbH (A Subsidiary of Merck KGaA, Darmstadt, Germany), Industriestrasse 25, 9471 Buchs, Switzerland
e-mail: christine.hellriegel@merckgroup.com

A. Rueck
e-mail: alexander.rueck@merckgroup.com

K. Breitruck
e-mail: kathrin.breitruck@merckgroup.com

M. Obkircher
e-mail: markus.obkircher@merckgroup.com

S. V. Medvedevskikh et al. (eds.), *Reference Materials in Measurement and Technology*,
https://doi.org/10.1007/978-3-031-06285-8_6

Introduction

In recent years, quantitative Nuclear Magnetic Resonance (qNMR) has become widely accepted as a very efficient and precise method for the quantification of organic compounds in many fields. Quantitative NMR has several outstanding advantages compared to other analytical techniques regarding purity determination, [4] especially the indifference towards impurities. An analyte can be measured directly against a reference and impurities do not affect the result as long as one single signal from sample and reference is pure enough for integration. This also includes non-detectable impurities such as salts and residual water. Jancke et al. proposed qNMR spectroscopy as a relative primary analytical method for measurement results at the highest metrological level since it can be described completely by mathematical equations [2, 7, 8].

The signal integrals are in direct proportionality with the number of protons or other nuclei contributing to the resonances [1]. Hereby, the structures of the chemical substances are fully irrelevant. Hence, a direct metrological traceability to the "système international "(SI) through internationally accepted primary reference standards can be achieved by comparing signal intensities of analyte and reference substance, which is usually not possible with chromatographic techniques [3, 9]. Since qNMR is increasingly used in the pharmaceutical industry and analytical testing laboratories, there is a growing demand for certified reference materials (CRM) for applications in different markets and fields. Sigma-Aldrich Production GmbH is holding an ISO/IEC 17025 [10] and ISO 17034 [11] accreditation as a manufacturer of certified reference materials for quantitative NMR since 2009. Through an accredited certification process, several CRM for the intended use in qNMR as well as for chromatographic applications were generated. In this workflow, various preliminary tests are first carried out since it has to be ensured that internal reference and analyte do not react and that peaks used for integration do not overlap. Subsequently, high-precision weighing is carried out, followed by the actual qNMR measurement, determination of the homogeneity, and setup of a full uncertainty budget. For these measurements, internationally accepted references (for example NIST SRM), which ensure metrological traceability to the SI, are crucial. At present, there are primary calibrants available from the National Metrology Institute of Japan (NMIJ)[1] and the National Institute of Standards and Technology (NIST),[2] which are designed for the intended use in qNMR. For one of those materials, NIST collaborated with Sigma Aldrich Production GmbH and the National Research Council Canada (NRC), to produce and certify

[1] NMIJ CRM 4601-b No. +++ 3.5-Bis(trifluoromethyl)benzoic Acid for Quantitative NMR (^{1}H, ^{19}F). https://unit.aist.go.jp/nmij/english/refmate/crm/cert/4601b_en.pdf.

NMIJ CRM 4602-A No. +++ 1.4-Bis(trimethylsilyl)-2,3,5,6-tetrafluorobenzene for Quantitative NMR (^{1}H, ^{19}F). https://unit.aist.go.jp/nmij/english/refmate/crm/cert/4602a_en.pdf accessed May 25th, 2021.

[2] NIST PS1 Primary Standard for quantitative NMR (Benzoic Acid). Certificate of analysis. https://www.nist.gov/system/files/documents/2020/07/16/NIST_PS1_COA.pdf accessed May 25th, 2021.

ultra-pure benzoic acid (PS1) as primary material with metrological traceability to the SI [5].

After the measurements for the certified value and homogeneity are completed, stability studies, data analysis and calculation as well as further characterization of the material are carried out. Finally, a certificate is issued which must contain a statement on metrological traceability guaranteeing the link to the SI unit as well as an expiry date, which indicates the period in which the certified value is valid within the stated expanded overall uncertainty. By following this workflow, multiple CRM neat materials with diverse solubilities and signals in different regions of the spectrum were developed, by using three NMR active nuclei ^{1}H, ^{31}P, and ^{19}F [12]. Those CRM can be used in testing or QC laboratories either for performing qNMR or in chromatographic analytical measurements. For chromatographic techniques, a broad range of CRM from different fields were developed, e.g. pesticides, extractables & leachables, flavors & fragrances, amino acids, antibiotics, fatty acid methyl esters (FAME), organic pollutants, polyaromatic hydrocarbons (PAH), phthalates, substances of very high concern (SVHC), toxins, vitamins, and many more. There is also an increasing interest in qNMR and demand for qNMR standards in the pharmaceutical industry, as the technique offers a straightforward alternative to the time-consuming mass balance approach for purity determination, but also for comparison to mass balance results, or to complement other methods [13, 14]. Compounds certified in this segment comprise Active Pharmaceutical Ingredients (API) and pharmaceutical impurities.

The large range of CRM for chromatographic applications is important, since in HPLC or GC scientists have to calibrate with an identical reference in order to receive quantitative results. In contrast, qNMR laboratories cannot use the same molecule as reference and analyte due to signal overlap, at least when using the internal quantification method. Therefore, a small set of CRM with a choice between different nuclei, chemical shifts and solubilities is sufficient for the qNMR user.

In addition to using CRM for quantification, it has to be demonstrated especially in regulated environments that the NMR instrument itself performs according to requirements given by the authorities or guidelines. After setup of the instrument by the supplier, typically an installation qualification (IQ) and operational qualification (OQ) is performed. While IQ evaluates the means of accommodating new equipment and testing its material, OQ is essential in challenging equipment parameters. Finally, Performance Qualification (PQ) is used to verify that the instrument is running according to the specification under real conditions. Users must perform PQ of their instrument, as directed by various authorities. A popular way is using a standard tube for NMR referencing, as it is regarded as supplying a particularly relevant test for the samples that are typically analyzed, in addition to standard NMR tests like line shape or ^{1}H sensitivity based on certified reference samples.

Most of our CRM qNMR standards have so far been provided in neat form to offer a maximum of flexibility. However, to facilitate the sample preparation and add convenience for the CRM user, a series of CRM solutions have been developed. Because for this type of CRM, the time-consuming dissolution step and part of the weighing is already done, the user is able to save laboratory time. The available CRM solutions comprise some of the most used calibration standards for ^{1}H and

^{31}P quantitative NMR dissolved in a deuterated solvent and filled into ampules. The certification procedure of qNMR CRM in deuterated solution is described here in detail.

Materials and Methods

The certification procedure for CRM solutions in amber glass ampules starts with the selection of highly pure starting material, which usually is well-characterized neat CRM. In this case, *Trace*Cert® materials from Sigma-Aldrich were used: Phosphono acetic acid, 96708; Fumaric acid, 76635; Calcium formate, 03826; Benzoic acid, 06185; Dimethyl sulfone 41867, Dimethyl terephthalate, 07038; 1,2,4,5-Tetrachloro-3-nitrobenzene, 40384; 1,3,5-Trimethoxybenzene, 74599. The following deuterated solvents were used: D_2O, Sigma-Aldrich, 151882; $CDCl_3$, Sigma-Aldrich, 151823; DMSO-d_6, Armar AG, 015000.2050. For this purpose, several pretests must be done in order to exclude solutions which are not suitable as CRM for qNMR measurements, since they have to be stable over a defined period of time. The selected CRMs are dissolved in various deuterated solvents and NMR spectra are recorded. The solutions are then stored at different temperatures (4 °C - 45 °C) and further NMR spectra are recorded periodically to determine the most stable combination.

Then, a high-precision weighing step on an analytical balance (XP205DR, Mettler Toledo) is used to create the solution of the starting material in a deuterated solvent according to the target value. This gravimetric step is also performed under ISO/IEC 17025 and ensures metrological traceability to the SI. During each weighing step, climate conditions are tracked for subsequent air buoyancy correction. Afterwards, the solution is dispensed in ampules in a defined and validated process under inert gas following the guidelines of ISO 17034 accreditation.

As outlined in ISO Guide 35:2017, [15] a sampling step is performed by taking representative ampules from the entire batch for homogeneity testing. In addition to the gravimetric value received through the weighing step, the certified value of the analyte in solution is determined by qNMR, as this is later the intended use.

For this qNMR measurement it is not only necessary to set the NMR parameters correctly, but important attention has also to be paid to the sample preparation. Of highest importance is the weighing step, because the weighing results directly influence the measured content and the measurement uncertainty. Therefore, the next steps in the quantification of the material are metrological weighing with an ultra-micro balance (XP6U or XPR6U, Mettler Toledo) of the solution and a suitable reference material including air buoyancy correction and subsequent qNMR measurements for the determination of the mass fraction.

All NMR experiments were performed on a Bruker Avance 600 MHz NMR instrument equipped with a 5 mm PABBO BB-^{1}H/D Z gradient probe. In most cases 16–64 scans were recorded for every sample with a ^{1}H relaxation time of $d1 = 60$ s and a ^{31}P relaxation time of 30 s, respectively. Quantification of the analyte content was directly calculated from the NMR peak areas, the number of nuclei, the molecular masses, the content of the reference material, and the weighing values of the analyte and reference substance. Those measurements are also used for the calculation of homogeneity of the material via analysis of variance (ANOVA) [15].

Extensive stability tests were taken into account during the certification process. [16] For this purpose, a representative amount of the finished ampules was used in accelerated stability tests (stress tests) at elevated temperature for a shorter time period as well as long-term stability tests at storage temperature for a longer time period. This step is followed by data analysis, uncertainty calculations, finalization of the certificate and review by quality assurance [17].

Metrological Traceability to the SI was achieved using primary reference material from the National Institute of Standards and Technology (NIST).

Results

By following the entire accredited process as outlined above, it was possible to certify suitable standards for intended use in ^{1}H or ^{31}P qNMR (see Fig. 1).

The overall uncertainty for the gravimetric preparation of these solution CRM are illustrated by the following cause-effect diagram in Fig. 2. It contains contributing factors from the weighing processes, the purity of the reference, as well as influencing factors from the ampule filling and homogeneity testing, which are determined and validated independently for each solvent.

The certified mass fraction values were determined by qNMR measurements. Calculation of the result is done by applying the following Eq. 1 [2]:

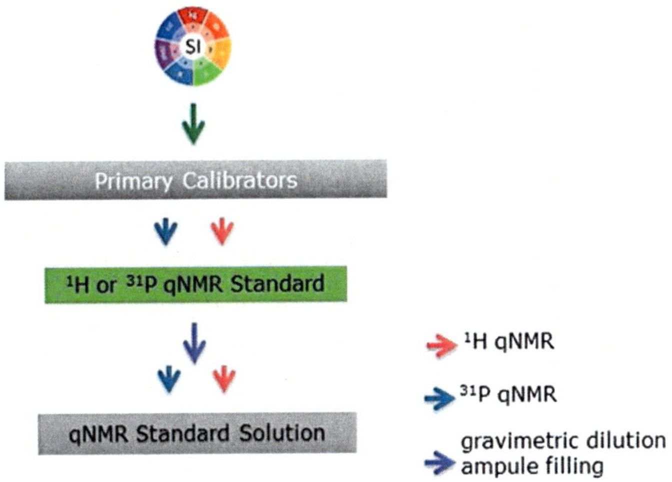

Fig. 1 Metrological traceability concept for qNMR standard solutions

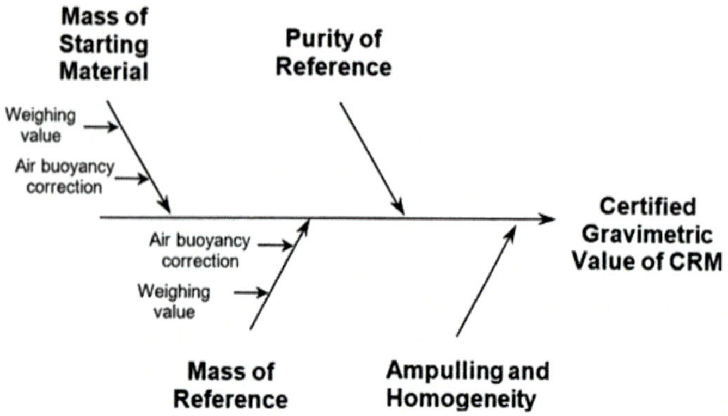

Fig. 2 Example Ishikawa diagram illustrating the potential sources of uncertainty in the reference value for a CRM solution, characterized by gravimetry

$$w_{sample}\left[\frac{mg}{g}\right] = \left(\frac{I_{sample}}{I_{Ref}} \times \frac{N_{Ref}}{N_{sample}} \times \frac{M_{sample}}{M_{Ref}} \times \frac{m_{Ref}}{m_{sample}} \times w_{Ref}\right) \times 1000 \tag{1}$$

w_{Sample}	mass fraction of the sample/[mg/g]	M_{sample}	Molecular mass of the sample/[g/mol]
w_{Ref}	mass fraction of the reference/[g/g]	M_{Ref}	Molecular mass of the reference/[g/mol]
I_{sample}	Integral of the sample signal/[-]	m_{Sample}	Mass of sample/[mg]
I_{Ref}	Integral of the reference signal/[-]	m_{Ref}	Mass of the reference/[mg]
N_{sample}	Number of nuclei generating the sample signal/[-]		
N_{Ref}	Number of nuclei generating the reference signal/[-]		

Additional factors need to be considered for the expanded uncertainty of the material, to fully comply with ISO/IEC 17025. The combined standard uncertainty is calculated by combination of the standard uncertainties of the input estimates according to Eurachem/CITAC Guide "Quantifying Uncertainty in Analytical Measurement." [18]

To fulfill the requirements of ISO 17034 [11], the full uncertainty budget includes in addition to the characterization measurement contributions from homogeneity assessment and stability studies, as mentioned before. The overall uncertainty can be described by the following Eq. (2) and results are shown in Table (1): [15]

$$u_{CRM} = \sqrt{u_{char}^2 + u_{hom}^2 + u_{lts}^2} \tag{2}$$

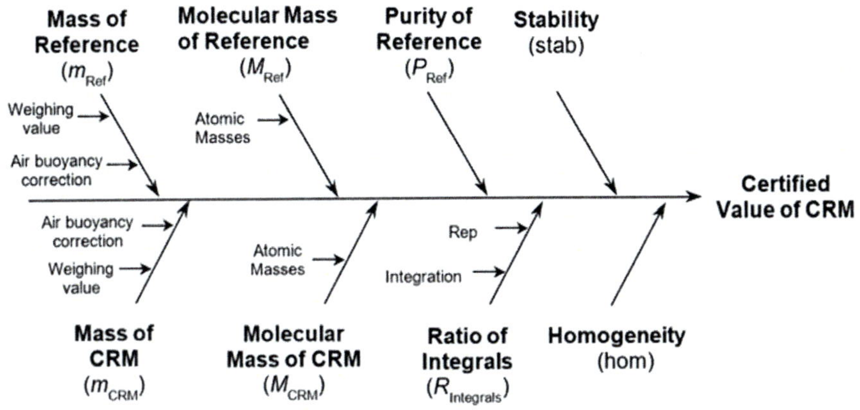

Fig. 3 Example Ishikawa diagram illustrating the potential sources of uncertainty in the reference value for a CRM solution, characterized by qNMR measurements

u_{CRM}	standard uncertainty associated with property value of the CRM
u_{char}	standard uncertainty associated with a value assigned in a characterization study
u_{hom}	standard uncertainty associated with heterogeneity
u_{lts}	standard uncertainty associated with long term stability

The certified value is ultimately given as mass fraction (mg/g) allowing the user to do gravimetric sample preparation, which gives the most accurate results. The uncertainty is calculated for a shelf life of 2 or 4 years, depending on the stability data. A certified value for the density is also provided on the certificate of analysis, to enable volumetric sampling as well if desired.

The uncertainty contributions for the qNMR certification of a solution CRM are illustrated by the following cause-effect diagram (Fig. 3):

Even though it is the value from the qNMR and not the gravimetric weighing measurement that appears in the certificate, the certified values derived from qNMR characterization are additionally compared to the gravimetric values, taking into account the uncertainties described above. The results were compared by applying the following Eq. (3) from ISO Guide 35 [15].

$$|x_{CRM} - x_{meas}| \leq k\sqrt{u_{CRM}^2 + u_{meas}^2} \tag{3}$$

x_{CRM}	property value of the CRM
x_{meas}	value from the qNMR measurement
u_{CRM}	standard uncertainty associated with property value of the CRM
u_{meas}	standard uncertainty associated with a value assigned by qNMR measurement

(continued)

(continued)

x_{CRM}	property value of the CRM
k	coverage factor ($k = 2$)

Table 2 shows the comparison of the value obtained during the production process (i.e. gravimetric step) with the certified value obtained by another measurement (i.e. qNMR), under consideration of the respective standard uncertainties and an appropriate coverage factor. The two results are matching with each other according to Eq. (3) for every single solution and thus the certified value can be considered confirmed. It is important to consider effects from ampule filling (for example possible evaporation of solvents, duration of the process especially with varying batch sizes) into the uncertainty budget, when determining a value by gravimetry. Those effects depend on the solvent and become more important with more volatile matrices.

The results of the determination of the mass fraction from qNMR experiments and gravimetry are illustrated in the following graphic (Fig. 4). The high volatility of deuterated chloroform is also illustrated by the higher level of uncertainty in the gravimetric production of the solution whereas the higher measurement uncertainty in the quantitative determination by means of qNMR of phosphono acetic acid can be attributed to the use of the method with ^{31}P. The ^{31}P 1D-NMR experiment is much less sensitive compared to the ^{1}H experiment, because of this the spectra are noisier and therefore a homogeneous evaluation of the experiments of the individual samples is more challenging resulting in higher measurement uncertainties.

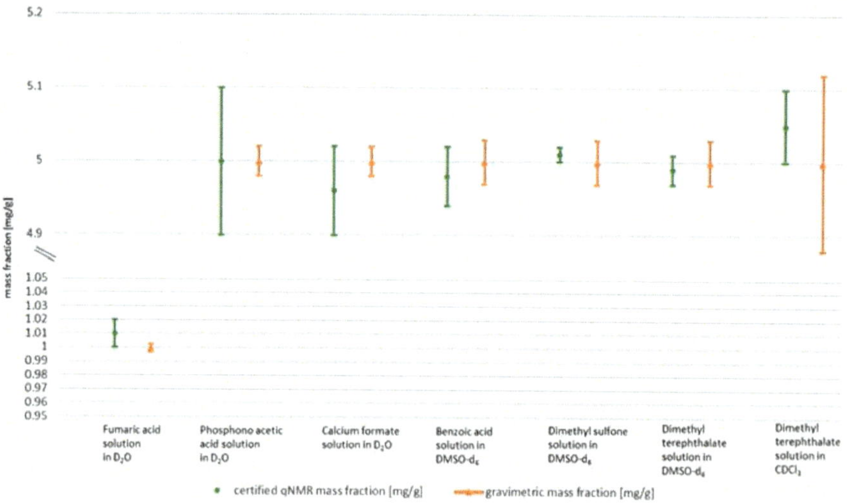

Fig. 4 Comparison of the mass fraction determined by qNMR (green) and by gravimetric production (orange) with their respective expanded measurement uncertainties. The scale had to be adjusted due to the lower mass fraction of the Fumaric acid solution

In order to demonstrate that instruments perform according to requirements given by the authorities, guidelines or internal SOPs, valid test samples have to be used.

There are clear advantages in using a performance qualification reference material that has been manufactured by a recognized high-quality manufacturer of such materials. This is partly because of convenience and speed. The ability to generate, test and certify high quality reference materials for the purposes described is not normally available in a typical NMR laboratory. Additionally, separation of the development process of the reference standard from the testing process provides a clear degree of independence.

A collaboration between an instrument manufacturer and Sigma-Aldrich Production GmbH has led to the development of certified reference material (CRM) tailored to quantitative performance qualification for NMR.

Even though the CRM for the quantitative Performance Qualification (qPQ) of NMR and potency determination has been developed using the same techniques that were already described above for the realization of the qNMR standard solutions by applying an internal standard certification method, there is one important difference. Since the end user must be able to directly assess the performance of the instrument, the CRM consists of a binary mixture of two well-characterized constituents. After a series of evaluations regarding applicability and stability 1,2,4,5-Tetrachloro-3-nitrobenzene (TCNB) and 1,3,5-Trimethoxybenzene (TMXB) were found to be ideal candidates in a DMSO-d_6 solution. By introducing a binary mixture, NMR users can compare the specific peaks of these two components against each other and use the mixture as a system suitability test.

The certification of this two-component mixture is an advancement of the process described for the single solution CRM. All steps that have already been described for the single solution CRM before, were also followed for the Performance Qualification mixture: characterization of the starting materials, choice of solvent, choice of separated peaks for quantification, determination of homogeneity and stability.

Equal to the qNMR CRM solutions described in Tables 1 and 2, the certified values of the qPQ CRM measured by qNMR can also be compared to the gravimetric results by applying Eq. (3). The results of the two different methods are shown in table 3 and show good correlation with each other, considering their respective uncertainties.

Contributions to the overall measurement uncertainty from the qNMR measurement, homogeneity and stability are in the same range as mentioned above (Table 2).

Discussion and Conclusion

The process of certifying neat materials by qNMR under ISO/IEC 17025 is well established at Sigma-Aldrich, Buchs, Switzerland for many years and in this paper, the successful certification of CRM solutions by qNMR was demonstrated, while

Table 1 Overview of the various CRM solutions including detailed uncertainty contributions from the qNMR measurement [u_{char}], homogeneity, and stability. An example of the influence of the various uncertainty contributions on the total measurement uncertainty was presented elsewhere [19]

Substance/ product no.*	Solvent	u_{char} [mg/g]	u_{hom} [mg/g]	u_{stab} [mg/g]	Certified qNMR mass fraction [mg/g] $\pm U$ ($k = 2$)
Fumaric acid solution/ 76201	D_2O	0.003	0.000	0.001	1.01 ± 0.01
Phosphono acetic acid solution/ 79251	D_2O	0.034	0.000	0.003	5.00 ± 0.10
Calcium formate solution/ 39606	D_2O	0.011	0.001	0.016	4.96 ± 0.06
Benzoic acid solution/ 39457	DMSO-d_6	0.009	0.000	0.016	4.98 ± 0.04
Dimethyl sulfone solution/ 16736	DMSO-d_6	0.017	0.016	0.004	5.01 ± 0.01
Dimethyl terephthalate solution/ 39387	DMSO-d_6	0.006	0.002	0.010	4.99 ± 0.02
Dimethyl terephthalate solution/ 39751	$CDCl_3$	0.021	0.012	0.003	5.05 ± 0.05

* For more information, please visit www.sigmaaldrich.com

complying to all steps mandatory for CRM production under ISO 17034 accreditation. Both types of qNMR CRM standards, neat material, and solution, have certain advantages depending on the application.

For the qNMR CRM in solution, it was demonstrated that metrological traceability can be achieved following two different routes, either through gravimetric dilution of well-characterized starting materials, or through qNMR measurements of the final product after the production process. Both routes are complying with the ISO 17034 guidelines and the respective results are in agreement with each other taking into account their corresponding measurement uncertainties.

The qNMR solution concept was in addition applied for a binary solution, containing two certified analytes. This qPQ (quantitative Performance Qualification) sample offers customers a ready-to-use solution to verify and control the performance of the NMR instrument, which is a crucial factor in order to receive reliable and accurate quantitation results.

Table 2 Key attributes and overview of the different qNMR solutions for ^{31}P and ^{1}H qNMR including chemical shift and individual relaxation time that was measured for each solution separately. The metrological traceability to the stated NIST SRM is realized either through a direct comparison or through an unbroken chain of measurements [19]

Substance/product no	Solvent	Intended use	Chemical shift δ [ppm]	Relaxation time T_1 [s]	Certified qNMR mass fraction [mg/g] $\pm U$ ($k = 2$)	Gravimetric mass fraction [mg/g] $\pm U$ ($k = 2$)	Metrological Traceability
Fumaric acid solution/ 76201	D_2O	^{1}H qNMR	6.5	12	1.01 ± 0.01	1.00 ± 0.01	NIST SRM 84 l (KHP) and NIST SRM 350b (Benzoic acid)
Phosphono acetic acid solution/ 79251	D_2O	^{31}P qNMR	15.7	5.3	5.00 ± 0.10	5.00 ± 0.02	NIST SRM 84 l (KHP)
Calcium formate solution/ 39606	D_2O	^{1}H qNMR	8.1	19.8	4.96 ± 0.06	5.00 ± 0.02	NIST SRM 84 l (KHP)
Benzoic acid solution/ 39457	DMSO-d_6	^{1}H qNMR	7.9 7.6 7.5	3.5 3.8 2.9	4.98 ± 0.04	5.00 ± 0.03	NIST PS1 (Benzoic acid)
Dimethyl sulfone solution/ 16736	DMSO-d_6	^{1}H qNMR	3	2.4	5.01 ± 0.01	5.00 ± 0.03	NIST PS1 (Benzoic acid)
Dimethyl terephthalate solution/ 39387	DMSO-d_6	^{1}H qNMR	8.1 3.9	3.4 1.2	4.99 ± 0.02	5.00 ± 0.03	NIST SRM 350b (Benzoic acid)
Dimethyl terephthalate solution/ 39751	$CDCl_3$	^{1}H qNMR	8.1 4	4.0 2.0	5.05 ± 0.05	5.00 ± 0.12	NIST SRM 350b (Benzoic acid)

Table 3 Key attributes and measurement values of the qPQ qNMR including chemical shift and individual relaxation time that was measured

Substance	Solvent	Intended use	chemical shift δ [ppm]	Relaxation time T_1 [s]	Certified qNMR mass fraction [mg/g] ± U (k = 2)	Gravimetric mass fraction [mg/g] ± U (k = 2)	Metrological Traceability
TCNB	DMSO-d_6	PQ	8.5	15	23.51 ± 0.18	23.50 ± 0.06	NIST PS1 (Benzoic acid)
TMXB	DMSO-d_6	PQ	6.1 3.7	3.5 1.5	5.00 ± 0.04	5.00 ± 0.03	NIST PS1 (Benzoic acid)

Acknowledgements The authors want to acknowledge the colleagues from the Sigma-Aldrich R&D team in Buchs, Switzerland, for their contribution to this field of work.

We want to thank Anna Codina, Christoph Freudenberger, Fabrice Moriaud and Ian Clegg from Bruker BioSpin for the successful collaboration in the field of performance qualification.

We also want to thank the reviewers of this paper for the willingness to take over this task.

Author Contributions All authors have contributed equally to the preparation of this article.

Conflict of Interest The article was prepared on the basis of a report presented at the IV International Scientific Conference "Reference materials in measurements and technologies" (St. Petersburg, December 1–3, 2020). The article was admitted for publication after the abstract was revised, the article was formalized and the review procedure was carried out.

References

1. Malz F, Jancke H (2005) Validation of quantitative NMR. J Pharm Biomed Anal 38(5):813–823. https://doi.org/10.1016/j.jpba.2005.01.043
2. Saito T, Ihara T, Koike M, Kinugasa S, Fujimine Y, Nose K, Hirai T (2009) A new traceability scheme for the development of international system-traceable persistent organic pollutant reference materials by quantitative nuclear magnetic resonance. Accred Qual Assur 14(2):79–86. https://doi.org/10.1007/s00769-008-0461-z
3. De Bièvre P, Dybkær R, Fajgelj A, Hibbert DB (2011) Metrological traceability of measurement results in chemistry: concepts and implementation (IUPAC Technical Report). Pure Appl Chem 83(10):1873–1935. https://doi.org/10.1351/PAC-REP-07-09-39
4. Weber M, Hellriegel C, Rück A et al (2013) Using high-performance quantitative NMR (HP-qNMR®) for certifying traceable and highly accurate purity values of organic reference materials with uncertainties <0.1 %. Accred Qual Assur 18:91–98. https://doi.org/10.1007/s00769-012-0944-9
5. Nelson MA, Waters JF, Toman B, Lang BE, Rück A, Breitruck K, Obkircher M, Windust A, Lippa KA (2018) A new realization of SI for organic chemical measurement: NIST PS1 primary standard for quantitative NMR (Benzoic Acid). Anal Chem 90(17):10510–10517. https://doi.org/10.1021/acs.analchem.8b02575
6. Westwood S, Yamazaki T, Huang T, Garrido B, Ün I, Zhang W, Martos G, Stoppacher N, Saito T (2019) Development and validation of a suite of standards for the purity assignment of organic compounds by quantitative NMR spectroscopy. Metrologia 56(6):064001. https://doi.org/10.1088/1681-7575/ab45cb
7. Jancke H (1998) NMR Spectroscopy as a primary analytical method. *Document 98/02 to the 4th Session of the CCQM* 1–12
8. Choi K, Myoung S, Seo Y, Ahn S (2021) Quantitative NMR as a versatile tool for the reference material preparation. Magnetochemistry 7(1):15. https://doi.org/10.3390/magnetochemistry7010015
9. Ellison SLR, Williams A (2019) Eurachem/CITAC Guide: Metrological Traceability in Analytical measurement: 2nd ed
10. ISO/IEC 17025 (2017) General requirements for the competence of testing and calibration laboratories. International Organization for Standardization, Geneva
11. ISO 17034 (2016) General requirements for the competence of reference material producers. International Organization for Standardization, Geneva
12. Rigger R, Rück A, Hellriegel C, Sauermoser R, Morf F, Breitruck K, Obkircher M (2017) Certified reference material for use in 1H, 31P, and 19F quantitative NMR, ensuring traceability to the international system of units. J AOAC Int 100(5):365–1375. https://doi.org/10.5740/jaoacint.17-0093

13. Davies SR, Jones K, Goldys A et al (2015) Purity assessment of organic calibration standards using a combination of quantitative NMR and mass balance. Anal Bioanal Chem 407(11):3103–3113. https://doi.org/10.1007/s00216-014-7893-6
14. Liu S-Y, Hu C-Q (2007) A comparative uncertainty study of the calibration of macrolide antibiotic reference standards using quantitative nuclear magnetic resonance and mass balance methods. Anal Chim Acta 602(1):114–112. https://doi.org/10.1016/j.aca.2007.09.009
15. ISO Guide 35 (2017) Reference materials- Guidance for characterization and assessment of homogeneity and stability
16. Weber M, Hellriegel C, Rueck A, Wuethrich J, Jenks P (2014) Using high-performance 1H NMR (HP-qNMR®) for the certification of organic reference materials under accreditation guidelines—describing the overall process with focus on homogeneity and stability assessment. J Pharm Biomed Anal 93(5):102–110. https://doi.org/10.1016/j.jpba.2013.09.007
17. ISO Guide 31 (2015) Reference materials-contents of certificates, labels and accompanying documentation
18. Ellison SLR, Williams A (2012) Eurachem/CITAC guide: quantifying uncertainty in analytical measurement, Third edition
19. Weber M, Hellriegel C, Rueck A, Wuethrich J, Jenks P, Obkircher M (2015) Method development in quantitative NMR towards metrologically traceable organic certified reference materials used as 31P qNMR standards. Anal Bioanal Chem 407:3115–3123. https://doi.org/10.1007/s00216-014-8306-6

Dr. Christine Hellriegel Senior Scientist R&D Reference Materials & Proficiency Testing; Sigma-Aldrich Production GmbH (A subsidiary of Merck KGaA, Darmstadt, Germany).

Dr. Alexander Rueck Manager R&D Reference Materials & Proficiency Testing; Sigma-Aldrich Production GmbH (A subsidiary of Merck KGaA, Darmstadt, Germany).

Dr. Kathrin Breitruck Senior Scientist R&D Reference Materials & Proficiency Testing; Sigma-Aldrich Production GmbH (A subsidiary of Merck KGaA, Darmstadt, Germany).

Dr. Markus Obkircher Director R&D Reference Materials & Proficiency Testing; Sigma-Aldrich Production GmbH (A subsidiary of Merck KGaA, Darmstadt, Germany).

Measurement Assurance of Phtalate Content: A Reference Material of Six Priority Phthalates in Methanol Solution

Anatoliy I. Krylov, Alena Yu. Mikheeva, Alexandra G. Budko, and Irina Yu. Tkachenko

Abstract The article provides information on the development of a new type of reference material (RM) of six priority phthalates (dimethyl phthalate, diethyl phthalate, di-n-butyl phthalate, benzyl butyl phthalate, di-2-ethyl hexyl phthalate, and di-n-octyl phthalate) in methanol solution, provided with metrological traceability to the GET 208–2019 (GET 208) national primary standard. The procedure for RM preparation and certification, including the homogeneity and stability assessment, is demonstrated. Pure organic substances (phthalates) characterised by the GET 208 primary standard were used as the starting material for the RM preparation using gravimetric and volumetric-gravimetric methods. The RM stability was assessed by an isochronous study. The mass fraction and mass concentration of individual phthalates in solution are the certified characteristics of the RM. The contributions from the purity of RM starting materials, preparation procedures, its heterogeneity and long-term instability were taken into account when calculating the uncertainty budget for the RM certified characteristics. The relative expanded uncertainty of certified values does not exceed 2%. As a result, the GSO (CRM) 11,366–2019 (6Ftlt-VNIIM) certified reference material of orthophthalic acid esters (phthalates) in methanol solution was developed and approved. The practical significance of using the GSO (CRM) 11,366–2019 consists in ensuring the metrological traceability of RMs to the corresponding SI units reproduced via the GET 208 national primary standard. The developed CRM can be used for different measuring tasks and metrological works.

A. I. Krylov (✉) · A. Yu. Mikheeva · A. G. Budko · I. Yu. Tkachenko
D. I. Mendeleyev Institute for Metrology (VNIIM), 19 Moskovskiy ave., Saint Petersburg 19005, Russia
e-mail: a.I.krylov@vniim.ru

A. Yu. Mikheeva
e-mail: a.mikheeva@vniim.ru

A. G. Budko
e-mail: aa@b10.vniim.ru

I. Yu. Tkachenko
e-mail: i.y.tkachenko@vniim.ru

S. V. Medvedevskikh et al. (eds.), *Reference Materials in Measurement and Technology*,
https://doi.org/10.1007/978-3-031-06285-8_7

Keywords Reference material · Phthalates · Certified value · Uncertainty · Metrological traceability · Mass balance method · Homogeneity · Stability · Measurement assurance

Abbreviations Used in the Article

CERI	Chemicals Evaluation and Research Institute, Japan
CIL	Cambridge Isotope Laboratories, Inc., USA
CRM	Certified reference material
NIM	National Institute of Metrology, China
NIST	National Institute of Standards and Technology, USA
NMIJ	National Metrology Institute of Japan
RM	Reference material
BBP	Benzyl butyl phthalate
HPLC–UV	High-performance liquid chromatography-ultraviolet detection
GSO	Formal name of certified reference material in Russia
GC–MS	Gas chromatography-mass spectrometry
DBP	Di-n-butyl phthalate
DMP	Dimethyl phthalate
DOP	Di-n-octyl phthalate
DEHP	Di-2-ethylhexyl phthalate
DEP	Diethyl phthalate
VOC	Volatile organic compound
MB	Mass balance
MI	Methodological instructions
NC	Non-volatile compounds
MC	Major component
PVC	Polyvinyl chloride
PRPS	Primary reference pure substance
RSC	Related structural compound
RM	Reference material
CRM	Certified reference material
TR CU	Technical regulations of the Customs Union
POS	Pure organic substance

Introduction

Russia produces a wide range of certified reference materials (CRMs) used for ensuring the uniformity of measurements [1]. However, there are particular groups (or classes) of chemical substances not provided with appropriate CRMs, which are in demand in various fields, including industry, agriculture, healthcare, ecology, etc. Another important question is the presence/absence of metrological traceability of reference materials (RMs), which determines their purpose and scope in analytical practice.

The esters of orthophthalic acid (phthalates) are large-scale products of the chemical industry, which are widely used as lubricating oils, anti-foaming agents, solvents, and carriers in various industrial processes, as well as in the cosmetics manufacture. Phthalates are most widely used as plasticisers in polyvinyl chloride (PVC) products obtained during the production of various polymeric materials for industrial, household, food, and medical purposes. The release of phthalates into environment is not only a result of their production and processing, but also a consequence of the widespread distribution of plastic in the form of consumer goods. In the bulk of a polymer, phthalates, as a rule, do not form strong bonds and are easily separated from finished products [2–5]. From the group of phthalates, six compounds are classified as priority organic pollutants: dimethyl phthalate (DMP), diethyl phthalate (DEP), di-n-butyl phthalate (DBP), benzyl butyl phthalate (BBP), di-2-ethyl hexyl phthalate (DEHP), and di-n-octyl phthalate (DOP) [6, 7].

Phthalates are low-toxic substances upon short-term exposure; however, under the conditions of long-term intake into the body, they can stimulate the occurrence of chronic diseases [8–12]. Based on scientific evidence from various sources, the European Union (EU) banned the use of DBP in the manufacture of cosmetics and toys in 2004 [13]. Later, the ban was extended over DEHP and BBP. Russia has also adopted a number of documents regulating the content of phthalates in packaging and products intended for children and adolescents, as well as in ambient air, textile, and shoe materials [13–19]. Methods for measuring the phthalate content in various matrices are prescribed in the Russian regulatory documents [20–26] and in the Technical Regulations of the Customs Union (CU).

Although phthalates belong to one of the most important groups of substances, until recently, no national (Russian) CRMs have been available for their measurement.

National metrological institutes in some countries and a number of commercial organisations produce a limited range of both certified reference materials (CRMs) and reference materials (RMs), information about which is summarized in Table. 1.

It is noteworthy that the assigned values in CRMs are characterised in terms of uncertainty (error) and established metrological traceability, which is the key difference between CRMs and RMs, for which these characteristics are not required [27].

A similar division on the basis of the presence/absence of metrological traceability of CRMs and RMs is recorded in the documents on standardization adopted by the Interstate Council for Standardization, Metrology, and Certification (Protocol No. 123-P of October 30, 2019) and put into effect in the Russian Federation [28].

The use of CRMs for performing measurements guarantees the highest accuracy and reliability, as well as the comparability of the results obtained, which is particularly important when performing measurements in areas subject to governmental

Table 1 Certified reference materials and reference materials for the phthalate group produced by foreign manufacturers

No	Manufacturer, country	Name of material	Reference number	Mass concentration[a], μg/cm³
Certified reference materials (CRM)				
1	NIM, China	DMP in methanol	GBW (E) 100,221	230 (±2%)
		DEP in methanol	GBW (E) 100,222	238 (±2%)
		DEHP in methanol	GBW (E) 100,223	186 (±2%)
		BBP in methanol	GBW (E) 100,224	165 (±2%)
		DBP in methanol	GBW (E) 100,226	195 (±2%)
2	CERI, Japan	DEP in methanol	CERIjcss-0569	1000 (±0.8%)
		DEP in hexane	CERIjcss-0570	1000 (±0.7%)
		DBP in methanol	CERIjcss-0571	1000 (±0.8%)
		DBP in hexane	CERIjcss-0572	1000 (±1.0%)
		DEHP in methanol	CERIjcss-0573	1000 (±0.7%)
		DEHP in hexane	CERIjcss-0574	1000 (±0.9%)
		BBP in methanol	CERIjcss-0575	1000 (±0.4%)
		BBP in hexane	CERIjcss-0576	1000 (±0.6%)
		8 Phthalates in Methanol: Diethyl phthalate Di-2-ethyl hexyl phthalate Di-n-butyl phthalate Benzyl butyl phthalate Di-n-hexyl phthalate Dicyclohexyl phthalate Di-n-pentyl phthalate Di-n-propyl phthalate	CERIjcss-0619	100 (±0.5%) 100 (±1.0%) 100 (±0.5%) 100 (±0.5%) 100 (±1.0%) 100 (±1.0%) 100 (±0.5%) 100 (±1.5%)
3	NIST, USA	6 Phthalates in Methanol: DMP DEP DBP BBP DEHP DOP	NIST SRM 3074	55.6 ± 1.2 51.4 ± 1.7 51.2 ± 1.2 52.2 ± 1.4 58.6 ± 1.3 48.2 ± 1.4
Reference materials (RM)				
4	CIL, USA	BBP in nonane	ULM-7551-1.2	100

(continued)

Table 1 (continued)

No	Manufacturer, country	Name of material	Reference number	Mass concentration[a], μg/cm^3
		DEHP in nonane	ULM-6241-1.2	1000
		DEP in nonane	ULM-6174-1.2	100
		DMP in nonane	ULM-6783-1.2	100
		DBP in nonane	ULM-7466-1.2	100
		DOP in nonane	ULM-6129-1.2	100
5	Supelco (Merck), USA	EPA method of phthalic acid ester mixture	48,805-U	2000 (each)
		DEHP in methanol	47,994	2000

[a]The values of relative expanded uncertainty are given in brackets

regulation, during the implementation of interstate projects, etc., when it is necessary to recognize the measurement results at the international level (CIPM MRA Agreement of October 10, 1999 [29]).

In this work, we aimed to develop a CRM for the composition of a solution of six priority phthalates, provided with metrological traceability to SI units and reproduced by the GET 208–2019 (hereinafter—GET 208) National Primary Standard of Mass (molar) Fraction and Mass (molar) Concentration of Organic Components in Liquid and Solid Substances and Materials based on liquid and gas chromatography–mass spectrometry with isotope dilution and gravimetry (Table 2).

Materials and Methods

Selection and Preparation of the RM Starting Material

In order to produce a phthalate CRM, pure organic substances (POS) of the following phthalates were purchased: DMP, DEP, DBP, BBP, DEHP, and DOP (Sigma-Aldrich, Germany) with the declared purity from 98.4% to 99.8% (Table 3).

The POS were studied using the analytical equipment included in the GET 208–2019.[1]

The first step in determining the purity of organic substances was to confirm their identity. Phthalates were identified by gas chromatography-mass spectrometry

[1] GET 208-2019 National primary standard of the units of mass (molar) fraction and mass (molar) concentration of organic components in liquid and solid substances and materials based on liquid and gas chromatography–mass spectrometry with isotope dilution and gravimetry.). In: Federal Informational Fund on Maintaining the Unity of Measurements. https://fgis.gost.ru/fundmetrology/registry/12/items/1382717.

Table 2 Operational conditions for the determination of related structural compounds (RSC) and volatile organic compounds (VOC) by GC–MS

Chromatograph	
Column	HP5-MS, 30 m × 0.25 mm ID × 0.25 μm df
Injector temperature	280 °C
Carrier Gas (Helium) Flow in the Constant Flow mode	1 cm^3/min
Carrier gas mode	Constant Flow
Sample injection mode	Split 1/50
Column thermostat temperature programme	40 °C (5 min) – 10 °C / min – 280 °C (35 min)
Delay for solvent exit	Without delay
Sample volume	1 mm^3
Mass Spectrometer	
Ion source temperature	230 °C
Quadrupole temperature	150 °C
Energy of ionising electrons	70 eV
Registration mode	Total ionic current (TIC) in the range m/z 33–550

(GC–MS) using the NIST 14[2] library of mass spectra and chromatographic retention indices.

At the second stage, the POS were thoroughly investigated for the presence of four probable groups of impurities (related structural compounds—RSC, water, volatile organic compounds—VOC, and non-volatile compounds—NC). Subsequently, the mass fraction of the major component was calculated according to the formula "100% minus the sum of impurities" [30–32] in compliance with the internationally recognized indirect method of mass balance (MB) for determining the purity of components.

The determination of RSC and VOC impurities was performed by GC–MS on an Agilent 7890B/5977B instrument (Agilent Technologies, USA).

The operating parameters of the chromatograph and mass spectrometer are presented in Table 2.

The determination of the mass fraction of NC impurities was carried out by the gravimetric method using the calibrated GH-252 (AND, Japan) electronic balance of a special accuracy class. The mass fraction of NC was measured by the difference in the mass of the flask before and after sample evaporation under reduced pressure ($T = 200$ °C, $P = 1.33$ kPa (10 mm Hg)), the result was related to the sample mass of 50 g.

[2] NIST Mass Spectral Library 2014.

Table 3 The content of major component (MC) in pure organic substances (POS) and certified characteristics of primary reference pure substances (PRPS)

Name of the component	Commercial Product	PRPS	
	Mass fraction of MC, %	Mass fraction of MC, %	Expanded uncertainty, %
Dimethyl phthalate	99.8	99.60	0.09
Diethyl phthalate	99.7	99.79	0.05
Di-n-butyl phthalate	99.7	99.46	0.17
Benzyl butyl phthalate	98.4	98.83	0.18
Di-2-ethyl hexyl phthalate	99.6	99.78	0.12
Di-n-octyl phthalate	98.6	99.35	0.18

The determination of the mass fraction of water was carried out on a Mettler Toledo C30 coulometric KF titrator (Mettler Toledo, Switzerland), which implemented the coulometric Karl Fischer titration, using the basic settings of the device.

The CRM material was prepared by gravimetric and volumetric-gravimetric methods.

Methanol previously checked for the residual content of counter impurities was used as a solvent for the preparation of a pilot batch of the CRM.

The weights of individual phthalates were determined on a calibrated XPE26 (Mettler Toledo, Switzerland) balance of the 1st (special) accuracy class. The control of blunders and the study of the CRM homogeneity and stability were carried out by GC–MS according to the phthalate content.

Results and Discussion

Fully characterised pure organic substances are a basis for ensuring traceability in organic analysis [33]. Thus, the first necessary step for the production of a CRM and the formation of a chain of metrological traceability is a detailed study of POS with the aim of its attestation/certification.

Determination of Purity of the Phthalate POS

The major component was identified by GC–MS using the NIST 14 library of mass spectra and chromatographic retention indices. The relevant and reference DBP mass spectra are shown in Fig. 1 as an example.

It should be noted that the ion with m/z = 149 is the main characteristic ion in the phthalate mass spectra with electron ionization, while the intensity of other ionic fragments is less than 20%, and the signal of the molecular ion is less than 10%. Thus,

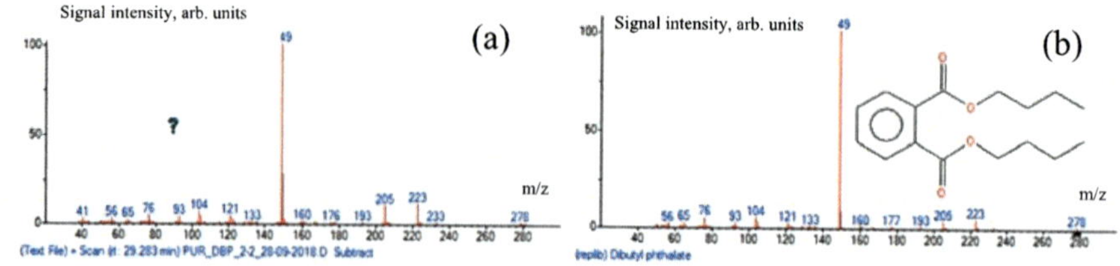

Fig. 1 Mass spectra: **a** relevant mass spectrum of the substance under study, **b** reference DBP mass spectrum from the NIST 14 mass spectral library

most phthalates have similar mass spectra, which makes their mass-spectrometric identification insufficient to confirm the identity of the major component. To increase the reliability of phthalate identification, we used the data on chromatographic retention indices. At least 80% coincidence of the obtained mass spectra with the reference ones and the deviation of retention indices not exceeding 5 units were taken as sufficient identification criteria.

Since phthalates are liquids with specified boiling points under normal conditions and, therefore, suitable for analysis by gas chromatography, the determination of RSC and VOC impurities was performed simultaneously and directly from the phthalate POS.

Identified RSC and VOC impurities were measured by an external standardization using point calibration characteristics under the assumption that the response factor of impurity relative to the corresponding standard is equal to 1. The following compounds were adopted as external standards:

- pentanol-1 for the impurities of aliphatic alcohols and ethers;
- acetic acid butyl ester for the impurities of aliphatic complex esters;
- benzoic acid for the impurities of aromatic carbonyl-containing compounds;
- 6 phthalates (DMP, DEP, DBP, BBP, DEHP, and DOP) for the impurities of phthalic acid ester.

In all investigated phthalate POS, characteristic impurities were registered, which presence was due to the production technology (alcohols, benzoic acid esters, phthalic anhydride, and isomeric phthalates). A typical mass chromatogram of phthalates, using DEHP as an example, is shown in Fig. 2.

As a result of the studies performed, it was found that the mass fraction of the sum of VOCs and RSCs in phthalates ranged from 0.17% to 1.12%.

The mass fraction of water was measured according to a conventional procedure using coulometric Karl Fischer titration. The content of water impurities in the investigated phthalates ranged from 0.027% to 0.105%.

To measure the content of non-volatile impurities, the gravimetric method was used, which consisted in measuring the mass of a sample before and after the evaporation of the major component (MC) and other volatile organic and inorganic substances. In all investigated phthalates, the mass fraction of impurities of non-volatile compounds was less than 0.0005%.

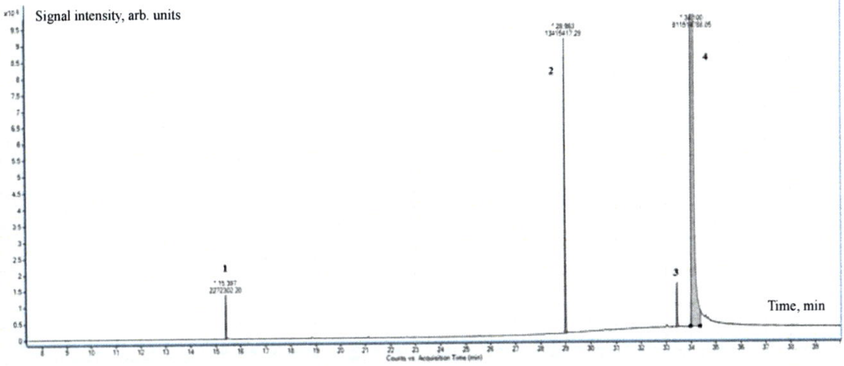

Fig. 2 Mass-chromatogram of DEHP: (1) 2-ethyl hexanol; (2) methyl-2-ethyl hexyl phthalate; (3) di-2-propyl pentyl phthalate; (4) DEHP

The generalized results of the measurements of the mass fraction of MC in POS and the certified characteristics of primary reference pure substances (PRPS) obtained on their basis are shown in Table 3.

Certification of the Phthalate RM in Solution

The value of the mass fraction and mass concentration of priority phthalates in the prepared solution (RM characterization) was established according to the calculation-experimental procedure of preparation.

The mass fraction of phthalates in the prepared solution (w_{an}, g/kg) was calculated taking into account the MC content in the corresponding PRPS (see Table 3) according to the general formula:

$$w_{an} = \frac{m_{ps} \cdot w_{ps}}{m_{sol} \cdot 100}, \tag{1}$$

where m_{ps} is the mass of the PRPS taken to prepare the solution, mg;

w_{ps} is the mass fraction of MC in the PRPS, %;

m_{sol} is the mass of solution, g.

The mass concentration of phthalates (ρ_{an}, g/dm^3) was calculated taking into account the MC content in the corresponding PRPS using the general formula:

$$\rho_{an} = \frac{m_{ps} \cdot w_{ps}}{V_{sol} \cdot 100}, \tag{2}$$

where V_{sol} is the volume of the prepared solution, cm^3.

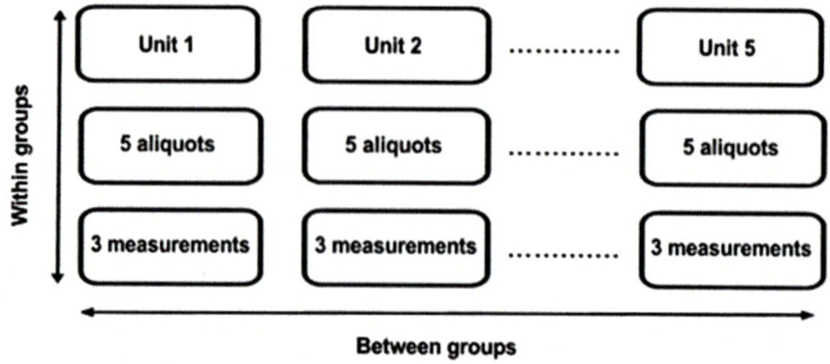

Fig. 3 Algorithm for studying the homogeneity of the RM for the composition of orthophthalic acid esters (phthalates) in methanol solution

RM homogeneity was evaluated by GC–MS. To this end, using the method of statistical selection, five specimens were selected from an experimental RM batch, and then from each specimen five aliquots of 0.25 cm^3 were taken. For each aliquot, three replicate determinations were carried out in accordance with the research algorithm shown in Fig. 3.

The obtained experimental data were processed in accordance with the ANOVA one-way analysis of variance, developed taking into account the provisions of [32, 34] and the methodology for studying the homogeneity and stability of CRMs given in the present paper. Calculations were performed for each phthalate under study.

No statistically significant alteration during the period of stability research was found. Based on the research results, it was concluded that the RM batch is homogeneous (for all phthalates $F < Fcr$). The contribution of uncertainty associated with material heterogeneity (u_h) was taken into account when calculating the uncertainty budget for the CRM characteristics (u_h was taken equal to 0.80%).

The RM stability was assessed during an isochronous study ("accelerated aging") [23]. Taking into account the recommended storage temperature range of the RM batch, the study of isochronous stability was carried out at a reference temperature of −18 °C and a test temperature of + 44 °C. The τ duration of the stability study (in days) was estimated by the formula:

$$\tau = \frac{T}{2^{\frac{t_1 - t_0}{10}}} \quad (3)$$

where T is the estimated shelf life of the RM unit, days;

t_0 and t_1 are the expected RM storage temperature of −18 °C and the storage temperature during accelerated aging of +44 °C, respectively.

Having fixed the values of storage temperature, and proceeding from the expected RM shelf life of at least 3 years, the τ duration of stability study was 14 days.

No statistically significant alteration was observed during the period of stability research. Based on the research results, the RM was acknowledged stable at the selected storage and transportation temperatures. The contribution of uncertainty associated with the RM instability (u_{stab}) was taken into account when calculating the uncertainty budget (u_{stab} was taken equal to 0.52%).

According to the research results, the shelf life of the RM was 3 years.

The calculation of the uncertainty budget of the CRM certified characteristics was carried out according to the general formula:

$$u = \sqrt{u_{char}^2 + u_h^2 + u_{stab}^2}, \tag{4}$$

where u_{char} is the relative standard uncertainty of the CRM certification method (%), calculated by formula (5):

$$u_{char} = \frac{u_{char}\left(w(or\rho)_{an}\right)}{w(or\rho)_{an}} = \sqrt{\left(\frac{u_{w_{ps}}}{w_{ps}} \cdot 100\right)^2 + \left(\frac{u_{m_{ps}}}{m_{ps}} \cdot 100\right)^2 + \left(\frac{u_{m(orV)_{sol}}}{m(orV)_{sol}} \cdot 100\right)^2} \tag{5}$$

u_h is the relative standard uncertainty resulting from the CRM inhomogeneity (%), calculated by formulas (6) or (7) in accordance with [34, 35]:

$$u_h = \frac{u_h\left(w(or\rho)_{an}\right)}{w(or\rho)_{an}} = \frac{\sqrt{\frac{(\overline{SS}_H - \overline{SS}_e)}{3} \cdot \frac{M(orV)_0}{M(orV)}}}{w(orC)_{an}} \cdot 100, \tag{6}$$

$$u_h = \frac{u_h\left(w(or\rho)_{an}\right)}{w(or\rho)_{an}} = \frac{\frac{1}{3}\sqrt{SS_e \cdot \frac{M(orV)_0}{M(orV)}}}{w(C)_{an}} \cdot 100 \tag{7}$$

u_{stab} is the relative standard uncertainty resulting from the CRM instability (%), calculated by formula (8) in accordance with [34, 35]:

$$u_{stab} = \frac{u_{stab}\left(w(or\rho)_{an}\right)}{w(or\rho)_{an}} = \frac{s_a \cdot t}{w(orC)_{an}} \cdot 100, \tag{8}$$

The relative expanded uncertainty (at k = 2) was calculated by the formula:

$$U\left(w(or\rho)_{an}\right) = 2 \cdot u, \tag{9}$$

The calculation results are summarized in Table 4.

Thus, the values of the CRM metrological characteristics were determined taking into account the results of:

- PRPS characterization;
- procedures for preparing the solution (according to the calculation-experimental procedure);
- studies of the CRM homogeneity;
- studies of the CRM stability.

Table 4 Uncertainty budget of the certified characteristics of phthalates in the CRM solution (mass fraction/mass concentration of the component)

NNo	Source of uncertainty	AAssessment type	Relative standard uncertainty, (contribution) % (Mass fraction)	Relative standard uncertainty, (contribution) % (Mass concentration)
1	Purity of PRPS	B	0.09	0.09
2	PRPS mass	B	0.0080	0.0080
3	Solution mass	B	0.010	–
	Solution volume	B	–	0.060
4	Inhomogeneity of RM	A	0.80	0.80
5	Instability of RM	A	0.52	0.52
Relative summarised standard uncertainty u			0.96	0.97
Relative expanded uncertainty ($k = 2$), U			1.92	1.94
Accepted			2	2

The metrological characteristics of the certified CRM batch are presented in Table 5.

As a result, the GSO (CRM) 11,366-2019 (6Ftlt-VNIIM) of orthophthalic acid esters (phthalates) in methanol solution was produced. The GSO (CRM) 11,366-2019 was used to calibrate analytical equipment in the development of a reference measurement procedure RMI VNIIM-243-02-2019 "Reference measurement procedure for

Table 5 Metrological characteristics of the CRM for the composition of phthalates in methanol solution

Name of substance (component)	Certified value of mass concentration of a component, mg/cm^3	Relative expanded uncertainty U at $k = 2^a$, %	Certified value of the mass fraction of a component, mg/g	Relative expanded uncertainty U at $k = 2^a$, %
DMP	2.04	2	2.58	2
DEP	2.05		2.59	
DBP	1.98		2.50	
BBP	1.98		2.51	
DEHP	2.02		2.55	
DOP	2.01		2.54	

[a]Corresponds to the limits of the permissible values of the relative error of the certified value of the CRM $\pm \delta$ (%) at $P = 0.95$

Table 6 RMSD of relative response factors obtained by HPLC–UV and GC–MS

Component	RMSD[a], %	
	GC–MS	HPLC–UV
DMP	0.57	0.12
DEP	0.55	0.10
DBP	0.80	0.11
BBP	0.80	0.11
DEHP	0.50	0.09
DOP	0.59	0.10

[a]Based on 15 measurements

the mass fraction of six priority phthalates (dimethyl phthalate, diethyl phthalate, di-n-butyl phthalate, benzyl butyl phthalate, di-2-ethyl hexyl phthalate and di-n-octyl phthalate) in objects based on polyvinyl chloride by gas chromatography–mass spectrometry with isotope dilution" [36].

The budget for the CRM uncertainty (see Table 4) shows that the contributions from the heterogeneity and instability of the material are the main components of the total standard uncertainty of the CRM certified values (up to 90%). We believe that this is due to the objective characteristics of the GC–MS method used for the study and does not indicate actual heterogeneity and/or instability of the CRM material.

To confirm this hypothesis, an experiment was designed using an alternative analytical method, high performance liquid chromatography with UV detection (HPLC–UV), which, in general, is characterized by a higher precision. The design of the experiment assumed an analysis of the aliquot of the CRM solution by the GC–MS and HPLC–UV methods under repeatability conditions. Based on the data obtained, the relative response factors of the analytes were calculated: DOP by DBP, other phthalates by DOP. The relative root-mean-square deviation (RMSD) of the measurement results obtained by different methods were determined (see Table 6).

Table 6 shows that, for the developed CRM, the contributions from inhomogeneity and instability significantly depend on the analytical method used (when using HPLC–UV, the RMSD of the measurement results improves by a factor of 5–7). Thus, if necessary, the accuracy characteristics of the GSO (CRM) 11,366–2019 can be noticeably improved by using an alternative procedure for assessing the homogeneity and/or stability of the CRM material.

Conclusion

As a result of the conducted studies, the GSO (CRM) 11,366-2019 (6Ftlt-VNIIM) for the composition of a solution of esters of orthophthalic acid (phthalates) in methanol was developed. The CRM is a solution of 6 individual phthalates in methanol packaged in (2.0 ± 0.1) cm^3 portions in hermetically sealed labelled glass ampoules with a nominal volume of 5 cm^3 and a shelf life of 3 years. The CRM certified characteristics are the mass fraction and mass concentration of individual phthalates (dimethyl phthalate, diethyl phthalate, di-n-butyl phthalate, benzyl butyl phthalate, di-2-ethyl hexyl phthalate, and di-n-octyl phthalate).

The CRM is provided with metrological traceability to the GET 208-2019 national primary standard, which guarantees the recognition of measurement results at the international level. The application of the GSO (CRM) 11,366-2019 may increase the accuracy and reliability of measurement results in solving any measuring tasks and performing various types of metrological work. Such work can include the development and certification of reference measurement procedures and techniques, organization of accuracy control of the measurement methods by the standard addition method and calibration of measuring instruments, interlaboratory comparative testing, etc.

Acknowledgements All measurements were carried out using the equipment of Mendeleyev Institute for Metrology (VNIIM).

Author Contributions A. I. Krylov: research supervision, concept and methodology development.

A. Yu. Mikheeva: analysis of literature data, design of experimental studies, critical analysis of experimental data, revision of the manuscript text.

A. G. Budko: collection of literature data, acquisition and analysis of experimental data, production of CRM, preparation of documents for CRM development and testing, manuscript draft preparation.

I. Yu. Tkachenko: collection of literature data, design of experimental studies, critical analysis of experimental data, revision of the manuscript text.

Conflict of Interest The article was prepared on the basis of a report presented at the IV International Scientific Conference "Reference Materials in Measurement and Technology" (St. Petersburg, December 1–3, 2020). The article was admitted for publication after the abstract was revised, the article was formalized and the review procedure was carried out.

The version in the Russian language is published in the journal "Measurement Standards. Reference Materials" 2021;17(3):5–19. https://doi.org/10.20915/2687-0886-2021-17-3-5-19.

References

1. Krylov AI, Mikheeva AY, Tkachenko IY (2017) Control of organic components: ensuring traceability of measurement results. Product Quality Control 11:12–17 (in Russ)
2. Rusakov PV (1988) Polymer production. Higher school, Moscow, 218 p (in Russ)
3. Korte F, Bahadir M, Kline V et al (1996) Environmental Chemistry. World publ, Moscow, pp 314–332 (in Russ)
4. Sekizawa J, Dobson S, Touch III RJ (2003) Concise international chemical assessment document 52. Diethyl phthalate. http://www.inchem.org/documents/cicads/cicads/cicad52.htm
5. Heise S, Litz N (2004) Deskstudy phthalates. German Federal Environmental Agency. Berlin, Horizontal, Germany, 40p
6. U.S. Environmental Protection Agency (2012) Phthalates Action Plan (Revised 03/14/2012). https://www.epa.gov/sites/default/files/2015-09/documents/phthalates_actionplan_revised_2012-03-14.pdf

7. Accessing the Market: European Union Phthalate Regulations. 2019 April 8. https://www.sgs.com/en/news/2019/04/accessing-the-market-european-union-phthalate-regulations
8. Markova OL, Ganichev PA, Yeremin GB, Zaritskaya EV (2020) Phthalate migration from packing materials for bottled water. findings of international studies. Health is the basis of human potential: problems and ways to solve them 15(1):416–427 (in Russ)
9. Ganichev PA, Markova OL, Yeremin GB, Myasnikov IO (2020) Effect of phthalates on population health. brief literary review. Health is the basis of human potential: problems and ways to solve them 15(1):233–239 (in Russ)
10. Wormuth M, Scheringer M, Vollenweider M, Hungerbühler K (2006) What are the sources of exposure to eight frequently used phthalic acid esters in Europeans? Wiley-Blackwell: Risk Management & Insurance Review 26(3):803–824. https://doi.org/10.1111/j.1539-6924.2006.00770.x
11. Luo Q et al (2018) Migration and potential risk of trace phthalates in bottled water: A global situation. Water Res 147:362–372. https://doi.org/10.1016/j.watres.2018.10.002
12. Chen X et al (2014) Toxicity and estrogenic endocrine disrupting activity of phthalates and their mixtures. Int J Environ Res Public Health 11(3):3156–3168. https://doi.org/10.3390/ijerph110303156
13. On amendments to the technical regulations of the customs union "On the safety of perfumery and cosmetic products" (TR CU 009/2011): decision of the Council of the Eurasian Economic Commission of March 29, 2019 No. 32. https://docs.cntd.ru/document/554460497
14. TR CU 009/2011 Technical regulations of the customs union "On the safety of perfumery and cosmetic products. https://docs.eaeunion.org/docs/ru-ru/0025181/cuc_30092011_799
15. TR CU 007/2011 Technical regulations of the customs union "On the safety of products intended for children and adolescents. https://docs.eaeunion.org/docs/ru-ru/0124930/cuc_30092011_797
16. TR CU 005/2011 Technical regulations of the customs union "On the safety of packaging". https://docs.eaeunion.org/docs/ru-ru/0025191/cuc_02092011_769
17. TR CU 008/2011 Technical regulations of the customs union "On the safety of toys". https://docs.eaeunion.org/docs/ru-ru/0124939/cuc_30092011_798
18. TR CU 017/2011 Technical regulations of the customs union "On the safety of light industry products (as amended on August 9, 2016). http://docs.cntd.ru/document/902320564
19. GN 2.1.6.3492-17 Maximum permissible concentration (MPC) of pollutants in the air of urban and rural settlements: Hygienic standards, as amended, approved. Resolution of the Chief State Sanitary Doctor of the Russian Federation 05/31/2018 no. 37. Federal Center for Hygiene and Epidemiology of Rospotrebnadzor, Moscow, 2019, 55 p (In Russ)
20. MUK 4.1.3160-14 Measurement of mass concentrations of phthalates (dimethyl phthalate, diethyl phthalate, dibutyl phthalate, benzyl butyl phthalate, di (2-ethylhexyl) phthalate) in milk by high performance liquid chromatography: Methodical guidelines. Moscow, 2015, 21 p (In Russ)
21. Method 606: Phthalate ester. U.S. Environmental protection agency. 1984, 20 p
22. GOST 25779-90 (2018) Toys. General safety requirements and control methods. Standartinform, Moscow (In Russ)
23. MUK 4.1.738-99 (1999) Chromato-mass spectrometric determination of phthalates and organic acids in water. Ministry of Health of Russia, Moscow, 13p (In Russ)
24. MUK 4.1.614-96 (1997) Guidelines for the determination of diethyl phthalate in atmospheric air by high performance liquid chromatography. Ministry of Health of Russia, Moscow, 454p (In Russ)
25. MUK 4.1.3168-14 Gas chromatographic determination of dimethyl phthalate, dimethyl terephthalate, diethyl phthalate, dibutyl phthalate, butyl benzyl phthalate, bis (2-ethylhexyl) phthalate and dioctyl phthalate in atmospheric air, the air of the test chamber and enclosed spaces. Ministry of Health of Russia, Moscow, 2015, 30 p (In Russ)
26. MUK 4.1.610-96 (1997) Guidelines for the gas chromatographic determination of dimethyl isophthalate in atmospheric air. Ministry of Health of Russia, Moscow, 454p (In Russ)

27. GOST ISO Guide 34-2014 (2014) General requirements for the competence of reference material producers. Standartinform, Moscow, 40p (In Russ)
28. GOST 8.315-2019 (2019) State system for ensuring the uniformity of measurements. Reference materials of composition and properties of substances and materials. Basic principles. Standartinform, Moscow, 39p (in Russ)
29. Mutual recognition of national measurement standards and calibration and measurement certificates issued by national metrology institutes: CIPM MRA of October 14, 1999. https://fgis.gost.ru/fundmetrology/registry/7
30. RMG 55-2003 GSI. State system for ensuring the uniformity of measurements. Reference materials for composition of pure organic substances (2004) Methods of certification. Principles basic. Publishing house of standards, Moscow, 12p (in Russ)
31. Duewer DL et al (2004) An approach to the metrologically sound traceable assessment of the chemical purity of organic reference materials. NIST, USA, p 47
32. GOST ISO Guide 35-2015 (2016) Reference materials. General and statistical principles for certification. Standartinform, Moscow, 61p (in Russ)
33. CCQM strategy document 2021–2030. Version 1.0 21.06.2021. https://www.bipm.org/documents/20126/41532413/CCQM+Strategy/31283069-94f4-f2c7-bbfc-7d652c9b3de8
34. RMG 93-2015 GSI (2011) State system for ensuring the uniformity of measurements. Estimation of metrological characteristics of reference materials. Standartinform, Moscow, 30p (in Russ)
35. R 50.2.058-2007 (2008) State system for ensuring the uniformity of measurements. Estimation of uncertainties of certified values of reference materials. Standartinform, Moscow, 30p (in Russ)
36. RMI VNIIM-243-01-2019 (2019) Reference method for measuring the mass fraction of six priority phthalates (dimethyl phthalate, diethyl phthalate, di (n-butyl) phthalate, benzyl butyl phthalate, di (2-ethylhexyl) phthalate and di (n-octyl) phthalate) in objects on based on polyvinyl chloride by gas chromatography/mass spectrometry with isotopic dilution. https://fgis.gost.ru/fundmetrology/registry/8/items/834820 (in Russ)

Anatoliy I. Krylov Dr. Sci. (Chem.), Head of the Department of State Measurement Standards in the field of organic and inorganic analysis.

Alena Yu. Mikheeva Cand. Sci. (Chem.), Leading Researcher of the Department of State Measurement Standards in the field of organic and inorganic analysis.

Alexandra G. Budko Researcher of the Department of State Measurement Standards in the field of organic and inorganic analysis.

Irina Yu. Tkachenko Leading Engineer of the Department of State Measurement Standards in the field of organic and inorganic analysis.

Development of Certified Reference Materials for Ensuring the Traceability of Mechanical Stress Measurements in Acoustic Non-destructive Testing

Ilona N. Matveeva and Vladimir V. Tolmachev

Abstract The article concerns the development of certified reference materials (CRMs) for steel mechanical properties. At present, in mechanical stress measurements, traceability of measurement results is ensured by the standard of time, rather than by the standard of mechanical stress. Therefore, a systematic approach to ensuring the uniformity of mechanical stress measurements requires the development of a reference. The present study is focused on the development and certification of CRMs having a special shape, traceable to SI units of force and length, and intended for transferring a unit of mechanical stress to measuring instruments, where the time-of-flight method of acoustic non-destructive testing is realized on the basis of the acoustoelasticity phenomenon. The key stages of developing CRMs, including the selection of the initial material, performing a homogeneity study, carrying out experimental studies using a certified reference (1st echelon state standard of the force unit), as well as establishing, on the basis of these studies, metrological characteristics of the developed CRMs. As a result of pattern approval tests, the GSO 11,544–2020/11545–2020 set of CRMs for steel mechanical properties was approved. The certified characteristics of the CRMs in the set involve the proportional limit mechanical stress; proof strength, plastic extension 0.2% R_p; tensile strength R_m; modulus of elasticity E. For the CRMs, the relative error of certified values do not exceed 12% at a confidence level of 0.95. The authors believe that, when applied as a reference in the verification schedule, the GSO 11544–2020/11545–2020 CRM set of steel mechanical properties will provide metrological traceability of the results in measuring mechanical stresses by instruments implementing the time-of-flight method of acoustic non-destructive testing.

Keywords Mechanical properties of steel · Stress–strain state · Temporary method of acoustic type of nondestructive check · The phenomenon of acoustoelasticity · Proof strength · Plastic extension 0.2% · Tensile strength · Modulus of elasticity · Certified reference material

I. N. Matveeva (✉) · V. V. Tolmachev
UNIIM—Affiliated Branch of the D. I. Mendeleyev Institute for Metrology, Krasnoarmeyskaya st. 4, Yekaterinburg 620075, Russia
e-mail: sertif@uniim.ru

S. V. Medvedevskikh et al. (eds.), *Reference Materials in Measurement and Technology*,
https://doi.org/10.1007/978-3-031-06285-8_8

Abbreviations

CRM Certified Reference Material
SSS Stress–Strain State
NDT Non-Destructive Testing

Introduction

Acoustic NDT is widely applied at hazardous production facilities in the gas, oil, and petrochemical industries, as well as in mechanical and nuclear engineering, utilities, and other industries due to the need for safety analysis of the feasibility of operating metal structures (main pipelines, NPP units, pressure vessels, supporting building structures, etc.) [1–13].

The probability of plastic destruction in critical components of structures and equipment can be evaluated using devices implementing the time-of-flight method of the acoustic NDT and providing the measurement of the current mechanical stresses arising from technological influences.

In order to ensure the reliability of results obtained in measuring mechanical stresses, such devices should have a traceability to the standard of the corresponding physical quantity (Fig. 1). However, as shown by the analysis of the register of approved patterns of measuring instruments, the measured value in these devices is represented by the transmission time of various type (P and S) elastic waves excited

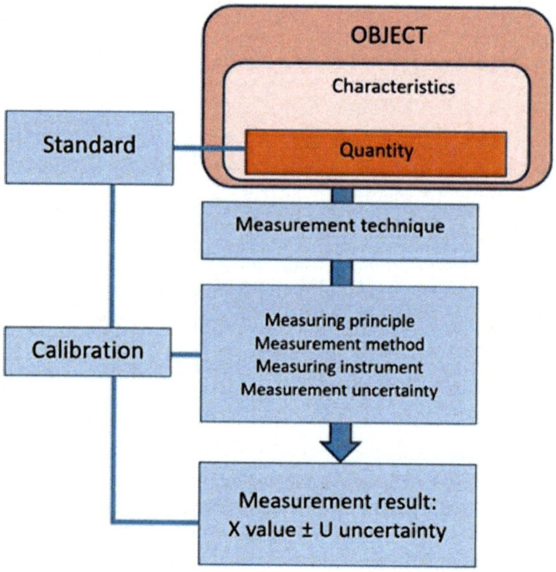

Fig. 1 Ensuring traceability of measurements to standards reproducing SI units

in the material of the controlled object, i.e., traceability is ensured to a time, not a mechanical stress, standard. The analysis of the verification procedures confirmed that the standards used for verification (oscilloscope, pulse generator, and frequency meter) do not transfer a unit of mechanical stress to the devices under verification.

In order to control SSS, standardized measurement techniques are used [14–18], which provide for the transfer of the dimension of a mechanical stress unit by constructing a calibration characteristic using a sample made of a material similar to that of the structure or equipment under study. At the same time, the fact that the properties of materials in the structure (equipment) under study were evaluated using samples of a different shape and dimensions is completely ignored, the contribution from the heterogeneity of the initial material in mechanical properties is not taken into account, and no uncertainty is estimated for the calibration characteristic obtained by the method of regression analysis. Thus, the basic principle of ensuring traceability in SSS measurements is violated: the result can be correlated with the reference system through a documented continuous chain of calibrations, each of which contributes to the measurement uncertainty [19].

Ensuring the traceability of mechanical stress measurements has two alternative perspectives.

The authors of [20] believe that, during the calibration of instruments implementing the time-of-flight method of the acoustic NDT, traceability to mass and length units is ensured, therefore, CRMs with an assigned value of the second kind modulus of elasticity (Young's modulus) can be applied for calibration.

The studies [21, 22] propose apply only CRMs for ensuring traceability to the mechanical stress unit, since obtaining a response from the stresses arising in the material during loading depends on external factors, i.e., the stresses are method-dependent and the traceability to the primary reference procedures of the National Metrology Institute is required.

Both methodological approaches recognize the use of CRMs as a necessary condition for ensuring the traceability of mechanical stresses.

Therefore, for mechanical stresses, the concept of traceability can be represented by Fig. 2, where the reference system consists of a material, sufficiently homogeneous and stable with respect to a certain property established in order to be used in the measurement process, i.e., a CRM [23].

The present study was aimed at the development and certification of special shape CRMs traceable to SI units of force and length and intended for transferring a unit of mechanical stress to measuring instruments, where a time-of-flight method of acoustic NDT is realized.

Materials and Methods

In order to ensure the representativeness of mechanical stress measurements in objects manufactured using various technological processing techniques, the CRM material was represented by:

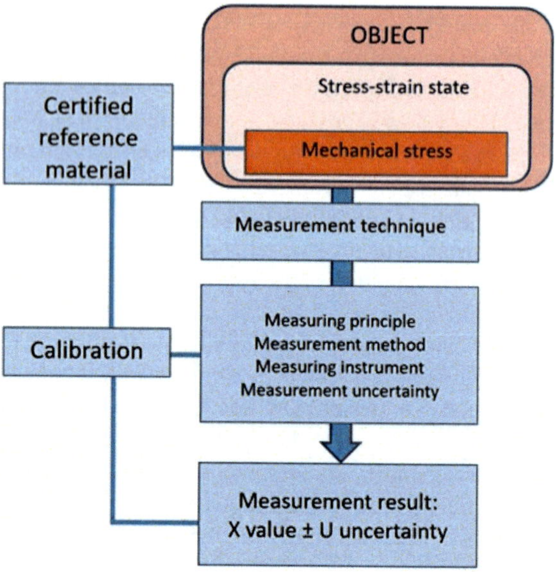

Fig. 2 Concept of traceability for mechanical stress

- rolled sheet made of structural steel;
- forging made of high-quality structural high-alloy steel.

The homogeneity study of the CRM in the form of a sheet or forging was a complex methodological task. Firstly, for calibration of measuring instruments implementing the time-of-flight method of the acoustic NDT, a CRM should have a shape and dimensions (hereinafter referred to as the form factor) different from proportional CRMs used to determine the mechanical properties under static tension [24]. Secondly, the main feature of measuring the mechanical properties of materials involves the fact that measurements cannot be repeated on the same sample due to its failure under loading.

The CRM inhomogeneity represents a source of a significant error (uncertainty) component; therefore, at the first stage, the acoustic anisotropy was estimated by an indirect method [18] in order to select an area, sufficient for the manufacture of a CRM on a sheet or forging according to the form factor and homogeneous in acoustic parameters. The results of evaluating the acoustic anisotropy are graphically represented in Fig. 3.

At the second stage, using proportional CRMs, the spatial homogeneity in the distribution of mechanical property was estimated in accordance with the algorithm provided in [25] by the method of static tension [24].

The CRMs were recognized as homogeneous. The obtained estimates of the error (uncertainty) from heterogeneity cannot be neglected; therefore, these values should be taken into account in estimating the error (uncertainty) of the CRM certified values.

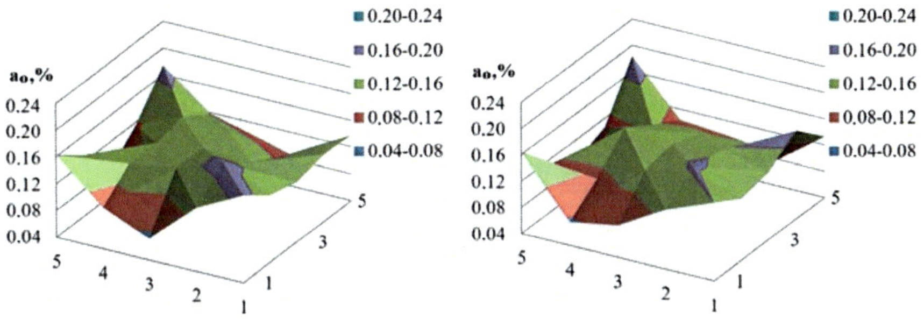

Fig. 3 Results of evaluating the acoustic anisotropy

The establishment of the CRM certified values, as well as the estimation of their errors, was carried out using the 1st echelon state standard of the force unit in accordance with the algorithms provided in [25].

The shelf life of the CRMs was established on the basis of the data about the shelf life for similar CRMs of the steel mechanical properties (GSO 10957–2017). The CRM shelf life comprises 10 years for all certified characteristics.

The developed CRM set was included in the State Register of approved patterns of certified reference materials as GSO 11544–2020/11545–2020.

The values of the GSO 11544–2020/11545–2020 metrological characteristics are presented in Tables 1 and 2.

The traceability of certified values to SI units was realized by means of direct measurements using the 1st echelon State standard of force unit in accordance with the State verification schedule for force measuring instruments and State verification schedule for length measuring instruments in the ranges from $1 \cdot 10^{-9}$ to 100 m and 0.2 to 50 μm (wavelengths).

CRMs are designed for pattern approval testing of instruments measuring mechanical stress using a time-of-flight method of acoustic non-destructive testing based

Table 1 Metrological characteristics of GSO 11544–2020

Certified characteristic	Unit designation	Interval of certified values	Absolute error of the certified value at a confidence level of 0.95, (± Δ)
Proportional limit[a]	MPa (N/mm^2)	100	± 12
Proof strength, plastic extension 0.2% R_p	MPa (N/mm^2)	166	± 12
Tensile strength R_m	MPa (N/mm^2)	307	± 12
Modulus of elasticity E	GPa	95	± 9

[a] The designation of the certified characteristics correspond to [24]

Table 2 Metrological characteristics of GSO 11545–2020

Certified characteristic	Unit designation	Interval of certified values	Absolute error of the certified value at a confidence level of 0.95, (± Δ)
Proportional limit[a]	MPa (N/mm^2)	830	± 19
Proof strength, plastic extension 0.2% R_p	MPa (N/mm^2)	1017	± 22
Tensile strength R_m	MPa (N/mm^2)	1198	± 12
Modulus of elasticity E	GPa	195	± 5

[a] The designation of the certified characteristics correspond to [24]

on the phenomenon of acoustoelasticity, verification and calibration of instruments measuring mechanical stress using a time-of-flight method of acoustic non-destructive testing, and the accuracy control in mechanical stress measurements.

Discussion and Conclusions

In the course of the study, unique CRMs of steel mechanical properties for ensuring the uniformity of mechanical stress measurements were developed and approved.

A distinctive feature of the developed CRMs involves the certification of the "proportional limit" and "modulus of elasticity E" characteristics. The certified values of these characteristics promoted for unambiguously setting the area of elastic deformation during verification and (or) calibration of the instruments measuring mechanical stress using the acoustoelasticity phenomenon, as well as for establishing the coefficients of elastic-acoustic coupling [16].

It should be noted that the certified values of the "modulus of elasticity E" characteristic differ from the reference data. The certified values of the characteristic for GSO 11544–2020, made of structural steel, and GSO 11545–2020, made of high quality structural high alloy steel, comprise 95 and 195 GPa, respectively, while their reference values are equal to 200 and 215 GPa. This difference is due to a number of reasons, each of which contributes to the reliability of the reference data. First, the available reference data contain no information on the uncertainty of the values obtained. Secondly, for reference data, no information is provided on the method of material processing for obtaining the reference data. For example, the CRM, made from rolled sheet, has surface hardening affecting the values of material properties obtained during testing. Therefore, the key task of ensuring the reliability of the results of mechanical property measurements involves the availability of the CRMs for calibration of non-destructive testing instruments associated with the technology of manufacturing controlled materials.

It should be noted that not only standards, but also reference systems in the form of GSO (State certified reference materials) and primary reference procedures should be used in verification schedules, representing the urgent task of legal metrology in the Russian Federation. The local verification schedule, based on the principles of Fig. 2, ensures using the GSO for pattern approval tests for instruments measuring mechanical stresses and implementing various methods of the acoustic NDT.

Acknowledgements The authors express their gratitude to the Deputy General Director for Development of INCOTES LLC Lev A. Pasmanik, the head of the scientific research sector in the field of measurements of force and mass of large values at D.I. Mendeleev Institute for Metrology, Ph.D., Ilya Y. Shmigelskiy for discussions during the development of certified reference materials, as well as reviewers for valuable comments on NDT terminology.

Author Contributions Ilona N. Matveeva: implementation of a formal analysis, writing a draft version of the article, creating visual materials.

Vladimir V. Tolmachev: development of the research concept, conducting research work, checking and editing the text of the article, management of research work.

Conflict of Interest The article was prepared on the basis of a report presented at the IV International Scientific Conference "Reference Materials in Measurement and Technology" (St. Petersburg, December 1–3, 2020). The article was admitted for publication after the abstract was revised, the article was formalized and the review procedure was carried out.

The version in the Russian language is published in the journal "Measurement Standards. Reference Materials" 2021;17(1).

References

1. Motova EA, Nikitina NY (2011) The possibility of ultrasonic testing of compressor blades after exploitation and repair. VESTNIK of Samara University. Aerospace Mechan Eng 10(3–2):52–56. https://doi.org/10.18287/2541-7533-2011-0-3-2(27)-52-56 (In Russ.)
2. Nikitina NE, Motova EA, Tarasenko YP (2012) Nondestructive testing of the working compressor blades of the aviation engine. VESTNIK of Samara University. Aerospace Mechan Eng 11(3–1):291–295. https://doi.org/10.18287/2541-7533-2012-0-3-1(34)-291-295 (In Russ.)
3. Nikitina NY, Kamyshev AV, Kazachek SV (2015) The application of the acoustoelasticity method for the determination of stresses in anisotropic pipe steels. Defektoskopiya 3:171–178 (In Russ.)
4. Nikitina NY, Kamyshev AV, Kazachek SV (2009) Using the phenomenon of acoustoelasticity in the study of the stress state of technological pipelines. Defektoskopiya 12:53–59 (In Russ.)
5. Nikitina NE, Kamyshev AV, Mironov NA (2009) Measurement of stresses in technological pipelines by the method of acoustoelasticity. Gazovaia promyshlennost' 5:64–67 (In Russ.)
6. Muravieva OV, Muraviev VV, Gabbasova MA, Petrov KV, Zorin VA (2016) Electromagnetic-acoustic structural analysis of rolled bars. AIP Conf Proc 1785(1):030017. https://doi.org/10.1063/1.4967038
7. Muravyev VV, Muravyeva OV, Strizhak VA, Pryakhin AV, Balobanov EN, Volkova LV (2011) Evaluation of residual stresses in rims of wagon wheels using the electromagnetic-acoustic method. Defektoskopiya 47(8):512–521 (In Russ.)
8. Motova EA, Nikitina NYe, Tarasenko Ur P (2016) Study of acoustic anisotropy of structural materials under variable loading. In: Problems and prospects for the development of engine building: materials of reports of the international scientific and technical conference, Samara, June 22–24 2016. Samara: Samara National Research University named after academician S.P. Queen, pp 16–17 (In Russ)

9. Mishakin VV, Gonchar AV, Knrashkin KV, Danilova NV (2009) The joint weld destruction study after static loading by acoustic method. Russian J Heavy Machin 7:27–30 (In Russ)
10. Belyaev AK, Polyanskiy VA, Grishchenko AI, Lobachev AM, Mansyrev DI, Modestov VS et al (2016) Application of the acoustic anisotropy approach for technical diagnostics of structures with large plastic deformations. AIP Conf Proc 1785(1):030004. https://doi.org/10.1063/1.4967025
11. Belyaev AK, Lobachev AM, Modestov VS et al (2016) Estimating the plastic strain with the use of acoustic anisotropy. Mech Solids 51:606–611. https://doi.org/10.3103/S0025654416050149
12. Grishchenko AI, Modestov VS, Polyanskiy VA, Tretyakov DA, Shtukin LV (2017) Experimental investigation of the acoustic anisotropy field in the sample with a stress concentrator. St. Petersburg Polytechn Univ J Phy Mathem 3(1):77–82. https://doi.org/10.1016/j.spjpm.2017.02.005
13. Alekseeva E, Alhimenko A, Belyaev A, Lobachev A, Polyanskiy V, Rostovykh G et al (2016) Evaluation of stress-strain state and cracking of weather-proof structural steel by acoustoelasticity. Constr Unique Build Struct 12(51):33–44 (In Russ.)
14. GOST R 52330–2005 (2005) Non-destructive testing. Evaluation of deformations in industrial and vehicle structures. General requirements. Standartinform, Moscow, p 5 (In Russ.)
15. GOST R 53966–2010 (2011) Non-destructive testing. Evaluation of mechanical stresses in constructions material. General requirements for the order of methods choice. Standartinform, Moscow, p 4 (In Russ.)
16. GOST R 55043–2012 (2014) Non-destructive testing. Evaluation of elastic-acoustic coefficients. General requirements. Standartinform, Moscow, p 8 (In Russ.)
17. GOST R 56664–2015 (2016) Non-destructive testing. Evaluation of stress state material engineering products by acoustoelastic methods. General requirements. Standartinform, Moscow, p 16 (In Russ.)
18. GOST R 52731–2007 (2007) Non-destructive testing. Stress evaluation by ultrasound. General requirements. Standartinform, Moscow, p 7(In Russ.)
19. RMG 29–2013 (2014) State system for ensuring the uniformity of measurements. Metrology. Basic terms and definitions. Standartinform, Moscow, p 60 (In Russ.)
20. Roebben G, Linsinger T, Lamberty A, Emons H (2010) Metrological traceability of the measured values of properties of engineering materials. Metrologia 47(2):23–31. https://doi.org/10.1088/0026-1394/47/2/S03
21. Bahng GW, Kim JJ, Lee HM, Huh YH (2010) Establishment of traceability in the measurement of the mechanical properties of materials. Metrologia 47(2):32–40
22. Czichos H, Saito T, Smith L (2011) Springer Handbook of Metrology and Testing. Springer, Berlin. https://doi.org/10.1007/978-3-642-16641-9
23. GOST ISO Guide 30–2019 (2019) Reference materials. Selected terms and definitions. Standartinform, Moscow, p 9 (In Russ.)
24. GOST 1497–84 Metals (1997) Methods of tension test. PPC Izdatel'stvo Standartov, Moscow, p 35 (In Russ.)
25. RMG 53–2002 (2004) State system for ensuring the uniformity of measurements. Reference materials. Evaluation of metrological characteristics with the use of measurement standards and reference devices. PPC Izdatel'stvo Standartov, Moscow, p 6 (In Russ.)

Ilona N. Matveeva Researcher of the Laboratory, UNIIM—Affiliated Branch of the D. I. Mendeleyev Institute for Metrology.

Vladimir V. Tolmachev Cand. Sci. (Phys.-Math.), Head of the Laboratory, UNIIM—Affiliated Branch of the D. I. Mendeleyev Institute for Metrology.

Application of Lyophilization for Preparing Reference Materials of Food Composition

Maria Yu. Medvedevskikh, Natalia L. Vostrikova, Anna S. Sergeeva, and Valeriya V. Studenok

Abstract This article describes the application of a lyophilization system for preparing reference materials (RMs) of food composition. The aim of the study was to develop a freeze-drying method for preparing an RM of poultry meat composition with certified values of mass fraction of moisture, nitrogen (protein), and fat. Poultry meat was used as the initial material: white (chicken breast, Sample 1) and red (chicken thigh, Sample 2). The procedure for preparing the RM included grinding, boiling, freezing, and lyophilization (syn. freeze-drying). Material homogeneity was studied using a standard air-heat drying technique from the State Primary Standard GET 173-2017. The mass fraction of nitrogen (protein) was measured using the State Secondary Standard GVET 176-1-2010. The mass fraction of fat was measured in accordance with the State Primary Reference Measurement Procedure. The total lyophilization time for Samples 1 and 2 was 19 and 28 h, respectively. The total weight loss was about 63% and 65% for Sample 1 and Sample 2, respectively. The difference in the values of moisture mass fraction for samples dried on different shelves of a freeze dryer was statistically significant, implying the heterogeneity of the material. A homogeneous material was obtained by additional homogenization procedures: grinding in a laboratory mill, sieving through a sieve, thorough mixing, and conditioning. The certified values of moisture, nitrogen, protein, and fat

---

The original version of this chapter was revised: The author names in Reference 10 have been updated. The correction to this chapter is available at https://doi.org/10.1007/978-3-031-06285-8_17

M. Yu. Medvedevskikh (✉) · A. S. Sergeeva · V. V. Studenok
UNIIM—Affiliated Branch of the D. I. Mendeleyev Institute for Metrology, Yekaterinburg, Russia
e-mail: lab241@uniim.ru

A. S. Sergeeva
e-mail: sergeevaas@uniim.ru

V. V. Studenok
e-mail: studenok@list.ru

N. L. Vostrikova
V. M. Gorbatov Federal Research Center for Food Systems of Russian Academy of Sciences, Moscow, Russia
e-mail: n.vostrikova@fncps.ru

S. V. Medvedevskikh et al. (eds.), *Reference Materials in Measurement and Technology*,
https://doi.org/10.1007/978-3-031-06285-8_9

mass fraction for Sample 1 were 4.5%, 14.74%, 92.1%, and 7.9%, respectively. For Sample 2, the respective values were 6.3%, 12.21%, 76.3%, and 23.8%. A lyophilization procedure was developed for producing an RM of the composition of boiled and freeze-dried poultry meat. This lyophilization system ensured the RM expiration period of 6 months at an ambient temperature of (7±3) °C and a relative humidity of no more than 60%. Following the results of this study, the developed RM of the composition of boiled and freeze-dried poultry meat was added to the Register of Certified RMs under the number GSO 11276-2019.

Keywords Food products · Reference materials · Standards · Primary reference measurement procedures · Lyophilization · Freeze drying · Nutritive values

Introduction

According to the ISO Guide 35 [1], water is the main interfering impurity in the development of reference materials (hereinafter RMs). An increased water content contributes to an increased heterogeneity of RMs, thereby reducing their expiration period. Therefore, creating RMs requires minimizing the water content. This is particularly important when developing RMs of food composition, since the mass fraction of water in the initial materials can exceed 80%. At the same time, it is important not only to remove the excess water, but also to preserve the structural integrity of the product, its protein composition, as well as the content of micro- and macroelements. This problem can be efficiently solved using lyophilization (*syn.* freeze-drying).

Lyophilization is a method for drying substances. Prior to drying, a liquid or solid substance undergoes deep freezing (at a temperature between −25 °C and −105 °C) followed vacuum and low- temperature treatment to remove the frozen solvent (sublimation). In other words, the solvent changes directly from solid to the gaseous state without passing through the liquid phase. This allows one: (1) to obtain dry cloths, medicine, products, etc. without losing their structural integrity and biological activity; (2) to reach high homogeneity values, thereby increasing the expiration period considerably [2].

Lyophilization is widely used for producing RMs of the composition of raw and processed foods. This accomplishes two tasks. First, the process increases the stability and extends the expiration period of RMs, for example, in the production of RMs of the composition of juice [3, 4] and meat products [5]. Second, lyophilization introduces additional components into the RM composition during the preparation process: vitamins, heavy metals, mycotoxins, polycyclic aromatic hydrocarbons (PAHs), pesticides, antibiotics, etc. [6–12]. However, until recently, this method of preparing RMs has not been used in the Russian Federation.

All the above mentioned explains the importance of developing lyophilization methods for producing homogeneous and stable RMs of food composition. This article studies the application of a lyophilization system for preparing RMs of food products using the example of RMs of the composition of poultry meat with the certified values of the mass fraction of moisture, nitrogen (protein), and fat.

Materials and methods

Poultry meat of two types was used as an initial material for preparing an RM: white (chicken breast, Sample 1) and red meat (chicken thigh, Sample 2). A freezing chamber LCexv 4010-23B-001 manufactured by Liebherr (Austria) and a freeze dryer Scientz-10ND manufactured by Scientz (China) were used for lyophilization. A Scientz-10ND dryer has three drying shelves (top, middle, and bottom) at different distances from the main unit, the drying conditions of which may vary. Table 1 shows the main technical parameters of the test equipment used for lyophilization.

Prior to lyophilization, the meat samples were boiled for one hour, parted from bones, and cut into pieces no more than 5 mm in size. The minced meat was spread in a thin layer onto three trays from the Scientz-10ND freeze dryer unit. The trays with meat were subsequently placed into an LCexv 4010-23B-001 freezer at a temperature of − 25 °C. At each preparation stage, the resulting material was weighed. The lyophilization process was performed in several stages (three or four depending on the weight of the initial material and the initial moisture content). Each stage included preliminary freezing in a freezer for at least 12 h and subsequent freeze drying. After the first stage, the material was ground in a laboratory mill Vyuga, produced by the Russian company EKAN, for 60 s at a speed of 12,500 rotations per minute, and then the ground material was placed into trays for drying. During lyophilization, a thermocouple was placed in the tray with the test material to monitor its temperature. The end of the lyophilization process was taken as the time when the sample reached a temperature of about 20 °C. After the last stage, the lyophilized meat was ground with a pestle in a mortar, sifted through a sieve with a nominal mesh size of 500 μm, thoroughly mixed, placed in a hermetically sealed container, and kept in a thermostat at a temperature of (7 ± 3)°C for 24 h.

The mass fraction of moisture in the dried material was measured by applying a standard air-heat drying unit for composition of the State Primary Standard of Units of Mass Fraction and Mass (Molar) Concentration of Water in the Solid and Liquid Substances and Materials GET 173-2017 [13]. The mass fraction of nitrogen (protein) was measured by the Kjeldahl method using the State Secondary Standard of Units of Mass Fraction and Mass (Molar) Concentration of Components in the Solid and

Table 1 Main technical parameters of the lyophilization equipment

Test equipment	Parameter	Value
Scientz-10ND	Capacitor temperature	up to − 55 °C (without loading)
	Maximum residual pressure	up to 10 kPa (without loading)
	Capacity	3 kg/24 h
LCexv 4010-23B-001	Operating temperature range	(7 … − 30)°C
	Temperature maintenance instability	± 0.6 °C
	Temperature deviation from the set one at − 25 °C	± 1.5 °C

Table 2 Changes in the mass of the RM at different preparation stages

No	Sample weight before boiling, g	Sample weight after boiling, g	Shelf no	Sample weight after boiling on each shelf of the freeze dryer, g
1	1,300	786.03	1	229.08
			2	276.98
			3	279.97
2	1,752	697.30	1	210.45
			2	233.15
			3	253.70

Liquid Substances and the Materials based on the Volumetric Titrimetric Method of Analysis GVET 176-1-2010 [14]. The measurements of the mass fraction of fat followed the State Primary Reference Measurement Procedure of a Mass Fraction of Fat in the Raw and Processed Food.[1]

Results and Discussion

Studying the Lyophilization Process When Preparing RMs

At each stage of preparing the RM, a decrease in its mass was observed (Table 2). As shown in Table 2, the weight loss when boiling was about 40% and 60% for Sample 1 and Sample 2, respectively.

During the lyophilization process, we observed a change in the temperature of the samples. The peculiarity of the analysis consists in the placement of a cooled sample (from the freezer) on the trays from the Scientz-10ND freeze dryer set at room temperature. In our experiment, the thermometer sensor was placed in the sample layer directly on the surface of the tray. As a result, at the beginning of the experiment, the maximum temperature gradient was observed at the boundary between the sample material and the tray. During the analysis, the tray and sample temperatures equalized. Hereinafter, this temperature identifies the temperature of the sample. Figure 1 shows the change in the temperature of Sample 2 (chicken thigh) at each stage of lyophilization.

The end of lyophilization was taken as the time, at which the sample reached a temperature of about 20°C. A further increase in temperature would have led to the

[1] The State Primary Reference Measurement Procedure of a Mass Fraction of Fat in the Raw and Processed Food M.241.01/RA.RU.311866/2018 (FR.PR1.31.2019.00001). In: Federal Informational Fund on Maintaining the Unity of Measurements. https://fgis.gost.ru/fundmetrology/registry/6/items/595556 (In Russian).

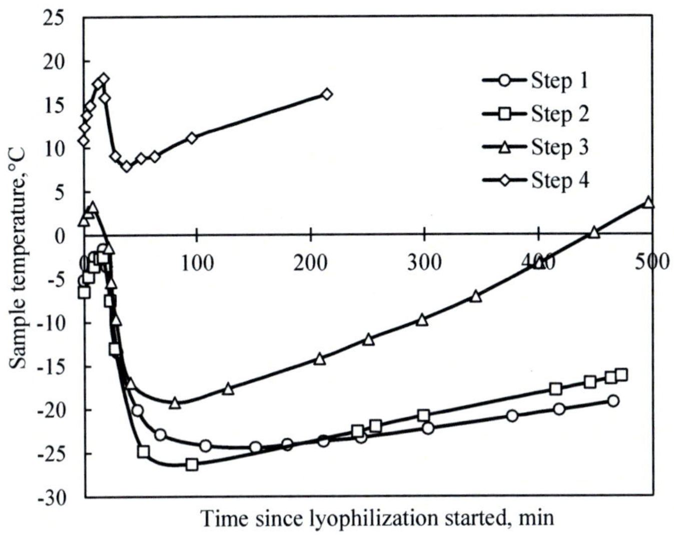

Fig. 1 The temperature change of Sample 2 (chicken thigh)

destruction of the material and the release of components other than water. The total lyophilization time for Samples 1 and 2 was 19 and 28 h, respectively.

The lyophilization process was accompanied by a significant weight loss of the samples. The results of calculating the weight loss of Samples 1 and 2 for each shelf during lyophilization are presented in Tables 3 and 4, respectively.

The total weight loss in the drying process was about 63% and 65% for Sample 1 and Sample 2, respectively.

Homogeneity of the Developed RM

The next stage was to assess the homogeneity of the material dried on different shelves in terms of the mass fraction of moisture using a standard unit, implementing the thermogravimetric method, from the State Primary Standard of Units of Mass Fraction and Mass (Molar) Concentration of Water in the Solids and Liquids and Materials GET 173-2017. Weighted portions ($n = 4$) were derived from the material dried on each shelf ($m = 3$), and the mass fraction of moisture was measured. The results of measurements of the mass fraction of moisture in Samples 1 and 2 are presented in Table 5.

The results presented in Table 5 were processed according to the algorithm described in [15].

The values

Table 3 Weight loss of Sample 1—poultry meat: white (breast)

Shelf no	Empty tray weight, g	Weight of the tray with the sample before lyophilization, g	Weight of the tray with the sample after lyophilization, g	Weight loss of the sample after lyophilization, %	Relative standard deviation of weight loss between the shelves, %
Stage 1					
1	242.67	467.65	372.30	42.38	9.6
2	245.37	518.05	422.10	35.19	
3	241.13	516.45	413.00	37.57	
Stage 2					
1	243.85	305.95	296.95	14.49	10.4
2	245.55	344.55	331.10	13.59	
3	242.30	315.45	303.30	16.61	
Stage 3					
1	243.85	296.95	284.55	23.35	22.6
2	245.55	331.10	299.85	36.53	
3	242.30	303.30	285.95	28.44	

$$W_i = \frac{1}{n_i} \sum_{j-1}^{n_i} Wij, \tag{1}$$

$$S_I^2 = \frac{1}{n_i - 1} \sum_{j=1}^{n_i} (Wij - Wi)^2, \tag{2}$$

were calculated for each $i = 1 \ldots m$;

W_i—the average value of moisture mass fraction of the i-group, %;

W_{ij}—the moisture mass fraction of the j-weighted portion from the i-group, %;

$S^2{}_i$—intragroup dispersions.

At the next stage, the hypothesis of equality of dispersions $S^2{}_i$ was tested by Cochran's Q test, for which the following value was calculated:

$$G_0 = \frac{S_{\max}^2}{\sum_{i=1}^{m} S_i^2}, \tag{3}$$

where

$S^2{}_{\max}$—the maximum value of intragroup dispersion;

$S^2{}_i$—intragroup dispersions.

If the following inequality was satisfied:

Table 4 Weight loss of Sample 2—poultry meat: red (thigh)

Shelf no	Empty tray weight, g	Weight of the tray with the sample before lyophilization, g	Weight of the tray with the sample after lyophilization, g	Weight loss of the sample after lyophilization, %	Relative standard deviation of weight loss between the shelves, %
Stage 1					
1	245.20	455.65	389.95	31.22	14.1
2	246.70	479.85	418.85	26.16	
3	243.65	497.35	437.05	23.77	
Stage 2					
1	245.00	386.95	339.70	33.29	7.5
2	246.00	414.70	362.25	31.09	
3	243.20	432.65	378.40	28.64	
Stage 3					
1	245.00	340.35	319.80	21.55	15.2
2	246.00	362.60	329.70	28.22	
3	243.20	378.55	339.85	28.59	
Stage 4					
1	245.00	320.75	319.25	1.98	67.4
2	246.00	330.40	326.65	4.44	
3	243.20	340.40	331.95	8.69	

Table 5 Measurements of the mass fraction of moisture in Samples 1 and 2 using GET 173–2017, %

Shelf no	Weighted portion no				Average	Relative standard deviation
	1	2	3	4		
Sample 1						
1	3.596	3.493	3.473	3.524	3.52	1.5
2	3.206	3.221	3.195	3.270	3.22	1.0
3	3.634	3.679	3.695	3.778	3.70	1.6
Sample 2						
1	5.125	5.117	5.106	5.155	5.13	0.4
2	5.321	5.289	5.413	5.416	5.36	1.2
3	5.488	5.553	5.531	5.542	5.53	0.5

$$G_0 \leq G_{\alpha,m,n-1}$$

where $G_{\alpha,m,n-1}$ is Cochran's Q test with the level of significance $\alpha = 0.05$ [16], the hypothesis of equality of dispersions is not rejected. If the hypothesis was not accepted, then a more thorough sampling was required.

If the hypothesis of equality of dispersion was accepted, the following values were calculated:

$$n = \sum_{i=1}^{m} n_i, \tag{4}$$

$$W = \frac{1}{n}\sum_{i=1}^{m}\sum_{j=1}^{n_i} Wij, \tag{5}$$

$$S_{01}^2 = \frac{1}{n-m}\sum_{i=1}^{m}(n_i - 1)S_i^2, \tag{6}$$

$$S_{02}^2 = \frac{1}{m-1}\sum_{i=1}^{m} n_i (Wi - W)^2, \tag{7}$$

$$F_o = \frac{S_{02}^2}{S_{01}^2}, \tag{8}$$

where

n—quantity of weighted portions in *m* groups;

W—the average value of moisture mass fraction for all groups, %;

S^2_{01}—average intragroup dispersion;

S^2_{02}—dispersion between the groups.

W_i and S^2_i are calculated following formulas (1) and (2), respectively. If the following inequality was satisfied:

$$F_0 \leq F_{\alpha,m-1,n-m} \tag{9}$$

where $F_{\alpha, m-1, n-m}$ is the quantile of F-distribution with $m - 1$ and $n - m$ degrees of freedom and the level of significance $\alpha = 0.05$ [16], the sample preparation of material for further studies was deemed satisfactory.

The results of assessing the material homogeneity for Samples 1 and 2 are presented in Table 6.

As Table 6 shows, the difference in the mass fraction of moisture of the material dried on different shelves was statistically significant; therefore, the material was

Table 6 Assessment of the material homogeneity in poultry meat: white (breast, Sample 1) and red (thigh, Sample 2)

Parameter	Parameter value	
	Sample 1	Sample 2
G_0	0.478	0.771
G(0.05;3;3)	0.798	0.798
W	3.48	5.34
$S^2{}_{01}$	0.0025	0.0018
$S^2{}_{02}$	0.2295	0.1641
F_0	90.097	91.029
F(0.05;3;9)	4.256	4.459

considered heterogeneous. The heterogeneity was reduced using additional homogenization procedures, including grinding with a pestle, sifting through a sieve, as well as thorough mixing, and conditioning.

Tests of the Developed RM

The resulting lyophilized and homogenized material was packaged into double sealed plastic bags 50 g each. To assess the certified value and heterogeneity characteristics of the developed RM, six RM copies were used. Each RM copy, randomly selected from the prepared batches, went through eight parallel screenings of the mass fraction of moisture and five parallel screenings of the mass fraction of nitrogen (protein) and fat. Processing the results of measurements, executed for the six RM copies to estimate the RM heterogeneity, was conducted according to single-factor variance analysis following the provisions of ISO Guide 35 [1]. A study of the RM stability at an ambient air temperature of (7 ± 3) °C and a relative humidity of no more than 60% was performed using the conventional method according to the ISO Guide 35 [1]. The method of regression analysis was used for processing the results. The expiration period of the RM was found to be six months.

The results of measuring the values of the certified values are represented in Table 7.

Table 7 Measurements of the mass fractions of moisture and nitrogen (protein) in boiled freeze-dried poultry meat

Indicator	Indicator value	
	Sample 1	Sample 2
Moisture content, %	4.5	6.3
Mass fraction of nitrogen*, %	14.74	12.21
Mass fraction of protein*, %	92.1	76.3
Mass fraction of fat*, %	7.9	23.8

* in terms of absolutely dry matter

Additional measurements of the values of the certified characteristics were performed using standardized measurement techniques [17] developed at the Gorbatov Federal Research Center for Food Systems of the Russian Academy of Sciences, Moscow. The consistency of the results obtained at the UNIIM and the Gorbatov Research Center confirmed the applicability of the developed RM for monitoring the accuracy of measuring the nutritional value of meat products.

Conclusion

Experimental research into the process of lyophilic drying as applied for preparing RMs was carried out. The aim was to evaluate the content of water (moisture) and heterogeneity of the RM prepared using a lyophilization system on a thermogravimetric unit of the GET 173-2017 composition.

A procedure of lyophilic drying for producing RMs of the composition of boiled and freeze-dried poultry meat with different moisture content and nitrogen (protein) values was developed. This lyophilization system ensured the RM expiration period of six months at an ambient temperature of $(7 \pm 3)°C$ and a relative humidity of no more than 60%. According to the obtained results, the RM of the composition of boiled and freeze-dried poultry meat was included in the register of approved types under GSO 11276–2019.[2]

Future studies into the use of lyophilization for preparing RMs should focus on the selection of lyophilization parameters for each specific substance.

Author Contributions M. Yu. Medvedevskikh: research conceptualization, methodology design, supervision over experimental studies.

N. L. Vostrikova: supervision over experimental studies, critical data analysis.

A. S. Sergeeva: literature review, original draft preparation, final text preparation.

V. V. Studenok: literature review, critical data analysis.

Conflict of Interest The article was prepared on the basis of a report presented at the IV International Scientific Conference "Reference Materials in Measurement and Technology" (St. Petersburg, December 1–3, 2020). The article was admitted for publication after the abstract was revised, the article was formalized and the review procedure was carried out.

The version in the Russian language is published in the journal "Measurement Standards. Reference Materials" 2021;17(1):35–45. https://doi.org/10.20915/2687-0886-2021-17-1-35-45.

[2] GSO 11274–2019 / GSO 11276–2019 The Standard model of the composition of the meat products of sublimation drying (collection MP-1 WITH UNIIM). In: Federal Informational Fund on Maintaining the Unity of Measurements. https://fgis.gost.ru/fundmetrology/registry/19/items/537358 (In Rus.).

References

1. ISO Guide 35:2017 Reference materials—Guidance for characterization and assessment of homogeneity and stability. ISO, Geneve
2. Panfilov VA (2009) Theoretical foundations of food technologies. In 2 books of book 2. Kolos, Moscow, 800p (In Russ.)
3. Ooghe W, Kramer GN, Schimmel H, Pauwels J (1998) Comparison of some additives used in the preparation of freeze-dried lemon juice candidate reference materials. Fresenius J Anal Chem 360(3–4):445–448. https://doi.org/10.1007/s002160050734
4. Kramer GN, Angelis LD, Pauwels J, Ooghe W, Belliardo JJ (1988) The freeze-drying of orange juice for the preparation of reference materials. Fresenius J Anal Chem 332:694–697. https://doi.org/10.1007/BF00472671
5. Ihnat M, Cloutier R, Wood D (1987) Reference materials for agricultural and food analysis: preparation and physical characterization of a bovine muscle powder candidate reference material. Fresenius' zeitschrift für analytische chemie 326:627–633. https://doi.org/10.1007/BF00473517
6. Bertolín JR, Joy M, Rufino-Moya PJ, Lobón S, Blanco M (2018) Simultaneous determination of carotenoids, tocopherols, retinol and cholesterol in ovine lyophilised samples of milk, meat, and liver and in unprocessed raw samples of fat. Food Chem 257:182–188. https://doi.org/10.1016/j.foodchem.2018.02.139
7. Carron P, Emteborg H, Hess P (2007) Freeze-drying for the stabilisation of shellfish toxins in mussel tissue (Mytilus edulis) reference materials. Anal Bioanal Chem 387(7):2475–2486. https://doi.org/10.1007/s00216-006-1104-z
8. Josephs R, Ulberth FE, Van Egmond HP, Emons H (2005) Aflatoxin M1 in milk powders: Processing, homogeneity and stability testing of certified reference materials. Food Addit Contam 22(9):864–874. https://doi.org/10.1080/02652030500016653
9. Huertas-Pérez JF, Bordajandi LR, Sejerøe-Olsen B, Emteborg H, Baù A, Schimmel H, Dabrio M (2015) Pahs in baby food: assessment of three different processing techniques for the preparation of reference materials. Anal Bioanal Chem 407:3069–3081. https://doi.org/10.1007/s00216-015-8490-z
10. Grimalt S, Harbeck S, Shegunova P, Seghers J, Sejerøe-Olsen B, Emteborg H, Dabrio M (2015) Development of a new cucumber reference material for pesticide residue analysis: feasibility study for material processing, homogeneity and stability assessment. Anal Bioanal Chem 407:3083–3091. https://doi.org/10.1007/s00216-015-8476-x
11. Zeleny R, Emteborg H, Charoud-Got J, Schimmel H, Nia Y, Mutel I, Ostyn A, Herbin S, Hennekinne J-A (2015) Development of a reference material for staphylococcus aureus enterotoxin A in cheese: feasibility study, processing, homogeneity and stability assessment. Food Chem 168:241–246. https://doi.org/10.1016/j.foodchem.2014.07.066
12. Won SH, Lee C-H, Kim B (2017) Development of certified reference materials for accurate determination of fluoroquinolone antibiotics in chicken meat. Food Chem 229:472–478. https://doi.org/10.1016/j.foodchem.2017.02.112
13. Medvedevskikh MYu, Sergeeva AS, Krasheninina MP, Shokhina OS (2019) Creating reference standards for the implementation of the State verification scheme for water content measurement. Meas Tech 62(6):475–483. https://doi.org/10.1007/s11018-019-01649-3(InRuss)
14. Krasheninina MP, Medvedevskikh MYu, Medvedevskikh SV, Sobina EP, Neudachina LK (2013) An estimate of the metrological characteristics of a standard sample of the composition of dried whole milk using primary and secondary state standards. Meas Tech 56(9):1076–1082. https://doi.org/10.1007/s11018-013-0333-8(InRuss)
15. Ivanov VP, Medvedevskikh SV (1988) Processing of the measurement results of water content in solid substances. Guidelines. Uro AN SSSR, Sverdlovsk, 200p (In Russ)
16. Bol'shev LN, Smirnov NV (1983) Tables of mathematical statistics. Science. Main editorial office of physical and mathematical literature, Moscow, 416p (In Russ)

17. Medvedevskikh MYu, Sergeeva AS, Krasheninina MP, Vostrikova NL, Semenova AA, Kuznetsova OA (2019) About the development of reference materials of meat and meat product composition. J Phys Conf Ser 1420:012030. https://doi.org/10.1088/1742-6596/1420/1/012030

Maria Yu. Medvedevskikh Cand. Sci. (Eng.), Head of the Laboratory for Metrological Support of Moisture Measurement and Reference Materials, UNIIM—Affiliated Branch of the D. I. Mendeleyev Institute for Metrology.

Natalia L. Vostrikova Dr. Sci. (Eng.), Head of the Research and Development Center, V.M. Gorbatov Federal Research Center for Food Systems of Russian Academy of Sciences.

Anna S. Sergeeva Cand. Sci. (Chem.), Senior Researcher of the Laboratory for Metrological Support of Moisture Measurement and Reference Materials, UNIIM—Affiliated Branch of the D. I. Mendeleyev Institute for Metrology.

Valeriya V. Studenok Engineer of the Department of the State Service of Reference Materials, UNIIM—Affiliated Branch of the D. I. Mendeleyev Institute for Metrology.

Metrological Maintenance of Gas Calorimeters and Wobbe Index Analyzers

Karina A. Mishina, Elena N. Korchagina, and Iaroslav V. Kazartsev

Abstract This article studies the issues of metrological maintenance of measuring instruments: gas calorimeters and Wobbe index analyzers. The purpose of this research lies in developing reference materials (RMs) of gases with the certified value of inferior calorific value, which is traceable to a state primary standard. The initial set of pure candidate gases (hydrogen, methane, ethane, and propane) and the target uncertainty of inferior calorific value (0.3%) were selected based on the results of analyzing the metrological characteristics of calorimetric equipment. The certified value of inferior calorific value is traceable to the State Primary Standard of the Energy Units of Combustion, Specific Energy of Combustion and Volumetric Energy of Combustion GET 16. The certified value for the selected gases and its uncertainty were evaluated by applying standard calorimeters-comparators for combusting the high and low-calorie gases USVG and USNG from GET 16. The results obtained during experimental studies and characterization of reference materials confirmed the declared accuracy values. Further studies should focus on developing reference materials of gas mixture-imitators for natural, associated, and other gases, as well as including the Wobbe index in the list of certified characteristics.

Keywords Inferior calorific value · Wobbe index · Gas calorimetry · Thermophysical properties · Reference material

K. A. Mishina (✉) · E. N. Korchagina · I. V. Kazartsev
D. I. Mendeleyev Institute for Metrology (VNIIM), 19 Moskovskiy ave., St. Petersburg, Saint Petersburg 190005, Russia
e-mail: k.a.mishina@vniim.ru

E. N. Korchagina
e-mail: e.n.korchagina@vniim.ru

I. V. Kazartsev
e-mail: y.v.kazartsev@vniim.ru

S. V. Medvedevskikh et al. (eds.), *Reference Materials in Measurement and Technology*,
https://doi.org/10.1007/978-3-031-06285-8_10

Abbreviations

GET 16	State Primary Standard of Units of Combustion Energy, Specific Energy of Combustion, and Volumetric Energy of Combustion GET 16–2018
GET 154	State Primary Standard of Units of Molar Fraction, Mass Fraction, and Mass Concentration of Components in Gas and Gas Condensate Media GET 154–2019
RMs	Reference Materials
MI	Measuring Instrument

Introduction

The efficient use of energy sources is attracting wide research interest due to aggravating environmental problems. In various areas of industrial production, the focus is made on the sustainable use of both traditional gas fuels and production by-products, such as coke-oven and blast furnace gases, associated petroleum gas, methane from gas mines, etc. Large-scale research in the field of creating renewable energy sources has brought about new methods for extracting low- and high-calorific combustible gases, including biogas, from agricultural, paper, and other industrial waste. The use of such gas fuels is currently growing at an average rate of 1.8% per year, and is projected to continue mainly due to an increasing demand in countries with stringent environmental regulations [1].

The listed types of gas have significant differences from natural gas in combustion energy and composition. Their composition and quality are subject to fluctuations both in the long and short term [2]. Therefore, when used as a fuel, special attention should be paid to the interchangeability of gases in order to avoid an increase in harmful emissions and damage to equipment during combustion, as well as to maintain the energy efficiency of production.

The main characteristic of the interchangeability of gas fuels—the Wobbe index—reflects information about the heat load in the system, since gases with different component composition and combustion energy, but the same Wobbe index, have the same thermal power and can be used without replacing burners.

The Wobbe index (higher W_{high} or lower W_{low}) is defined as the ratio of the combustion energy (higher H_{high} or lower H_{low}) of a gas to the square root of its relative density d in the air (Formula 1):

$$W_{\text{high/low}} = \frac{H_{\text{high/low}}}{\sqrt{d}}, \tag{1}$$

The relative density of gas in the air d can be calculated as the ratio of the absolute gas density ρ_g to the air density ρ_{air}, which is 1.204449 kg $\cdot$ m^{-3} under normal conditions (293.15 K, 101.325 kPa) [3].

Table 1 Maximum permissible measurement errors in accordance with ISO 15971:2008 [4]

MI class	0	1	2	3
H_{high}	± 0.1%	± 0.1 MJ/m^3 (~0.25%)	± 0.2 MJ/m^3 (~0.25%)	± 0.5 MJ/m^3 (~0.25%)

Table 2 Maximum permissible measurement errors in accordance with OIML R 140:2007 [6]

Precision class	A	B	C
Energy content	± 1.0%	± 2.0%	± 3.0%
H_{high}	± 0.5%	± 1.0%	± 1.0%

The net calorific value H_{low} can be calculated in accordance with GOST 31,369–2008, using a calculation method for determining the combustion energy and other gas properties on the basis of its component composition (determined with a gas chromatograph) [3], or determined experimentally by the calorimetric measurement methods described in ISO 15971:2008 [4]. Compared to the calculation according to GOST 31,369–2008, calorimetric measurement methods have a number of advantages when analyzing unconventional types of gas fuels: absence of limitations in the content of components (for example, ethane and hydrogen), lower uncertainty and calorimeter response time, as well as the efficiency and simplicity (in terms of maintenance) of calorimetric systems, including procedures of their calibration [5].

At the moment, various types of measuring equipment are used for determining the combustion energy and the Wobbe index, the operation principles of which rely on calorimetric methods. International standards ISO 15971:2008 and OIML R 140:2007 describe the classification and regulatory requirements for the accuracy of such equipment (Tables 1 and 2).

The quality of the results obtained with such equipment depends significantly on the correctness of its calibration. In the absence of reference materials (RMs), the ability to perform these routine procedures is severely limited. ISO 15971:2008 proposes pure methane, gas mixtures with a calculated combustion energy value, and gases certified with reference calorimeters as calibration standards. For the equipment used in a range other than natural gas, the former two options are not optimal. The latter option is preferable, since it can achieve a lower uncertainty of combustion energy measurements and ensure direct traceability of the measurement results to the primary standard.

In Europe, various aspects of gas research are handled by GERG (European Gas Research Group, Belgium), founded in 1961 and currently comprising 24 organizations from 10 European countries.

Several years ago, in order to ensure traceability of measurements in this area, GERG conducted a project aimed at the development of reference gas calorimeters with a measurement uncertainty not exceeding ± 0.05% at PTB (Germany) and LNE (France) [7, 8]. The results obtained using these calorimeters and the KATET calorimeter from GET 16–2018[1] made it possible to clarify the value of methane combustion energy in ISO 6976:2016 [9] and to reduce the uncertainty of this value [10–12].

Currently, most of the research in the field of combustion energy measurements is focused on biogas and other gases from unconventional sources [13–15]. The tasks of obtaining reliable and consistent values during measurements using working measuring instruments (MI) are also relevant. Among other research problems are the use of computational methods in gas transmission networks [16] and mixed flows [17] and comparison of various methods and instruments for measuring combustion energy of gases [4, 17, 18]. It should be noted that, at the moment, other countries lack RMs of volumetric combustion energy traceable to national standards.

In Russia, until 2021, the metrological maintenance of measuring instruments in the field of gas calorimetry had been carried out with the use of working standards developed at the D. I. Mendeleyev Institute for Metrology (VNIIM)[2] in St. Petersburg and used for verification and testing of gas calorimeters for type approval purposes. However, these tools did not satisfy the need for available calibration standards with certified combustion energy values, which are required during periodic routine maintenance of calorimetric equipment.

Therefore, the aim of this study was to develop a set of certified reference materials (RMs) of the volumetric inferior (lower) calorific value (LCV) based on pure gases—hydrogen, methane, ethane, and propane. The estimated uncertainty of the certified inferior calorific value for this set shall not exceed 0.3%.

Materials and Methods

To develop RMs for establishing and monitoring the metrological characteristics of gas calorimeters and Wobbe index analyzers, 4 samples of pure gases were selected and studied—hydrogen, methane, ethane, and propane. The requirements for the metrological characteristics of the developed RMs, as well as the choice of the samples under study, were determined by the results of analyzing the metrological characteristics of working measuring instruments. The certified inferior calorific value for the studied samples and its uncertainty were calculated based on the experimental data obtained by direct calorimetric method using the calorimeters-comparators for combustion of high- and low-calorie gases (USVG and USNG) from the State Primary Standard of the Combustion Energy Units GET 16–2018.

---

[1] GET 16-2018 State primary standard of units combustion energy, specific combustion energy and volumetric combustion energy. In: Federal Information Fund for Ensuring the Uniformity of Measurements. https://fgis.gost.ru/fundmetrology/registry/12/items/397905 (In Russ.).

[2] D. I. Mendeleyev Institute for Metrology (VNIIM). https://www.vniim.ru/etalon.html (In Russ.).

Selection and Preparation of Initial RMs

We took into account that the RMs of gases under development are intended for evaluation of metrological characteristics of calorimetric equipment during verification, calibration, and type approval tests.

Table 3 shows the main metrological characteristics of gas calorimeters and complex analyzers, designed for measuring combustion energy and the Wobbe index that passed the type approval tests in the Russian Federation.

An analysis of the characteristics of working measuring instruments showed that the most common are gas calorimeters and Wobbe index analyzers with a wide measurement range, accuracy class 3 according to ISO 15971:2008 (accuracy class B according to OIML R 140: 2007) with a tolerance limit of ± 1.0% and more.

Table 3 Main metrological characteristics of gas calorimeters operated in the Russian federation

Measuring instrument	Measured value	Measuring range	Limits of the permissible measurement error, %
9610, Cosa Xentaur Corporation	$W_{low/high}$, $H_{low/high}$	1–95 MJ/m^3	± 1.0[a]
	d	0.1–3.0	± 0.6
NKS, Teplofizichiskie pribory	H_{low}	25–41 MJ/m^3	± 0.3
RBM 2000, Reineke Mes-und Regeltechnik GmbH	$W_{low/high}$, $H_{low/high}$	3–100 MJ/m^3	± 1.0[a]
		28–40 MJ/m^3	± 0.6[a]
	ρ	0–2 kg/m^3	± 1.0[a]
CWD 2000, UNION Instruments GmbH	H_{low}	2.5–50.0 MJ/m^3	± 1.0[a]
	W_{low}	2.5–55.5 MJ/m^3	± 1.0[a]
CWD 2005, UNION Instruments GmbH	$H_{low/high}$	3.0–5.0 MJ/m^3	± 1.0[a]
		12.5–21.0 MJ/m^3	
WI, Reineke Mes-und Regeltechnik GmbH	H_{low}	2.5–55.5 MJ/m^3	± 1.5[a]
	W_{low}	2.5–56.0 MJ/m^3	± 1.0[a]
KSNG-05, Teplofizichiskie pribory	H_{low}	25–50 MJ/m^3	± 0.5
		31.8–39.8 MJ/m^3	± 0.45
EMC 500, RMG Messtechnik GmbH	$H_{low/high}$	30.2–47.2 MJ/m^3	± 0.5
	ρ	0.65–1.3 kg/m^3	± 0.5
66, Reineke Mes-und Regeltechnik GmbH	H_{high}	32–42 MJ/m^3	± 0.5[a]
WIM 9900, Horbe Instruments B.V	W_{low}	40–60 MJ/m^3	± 2.0
	d	0–2	± 0.2
WIM Compas	$W_{low/high}$	0–90 MJ/m^3	± 1.0[a]
	$H_{low/high}$		± 0.7[a]
	d	0.5–1.7	± 1.0[a]

[a]Limits of permissible measurement error, reduced to the upper limit of the measurement range

Table 4 Certified characteristics, the range of permissible certified values, and permissible values of the expanded uncertainty of the RM of the inferior calorific value of gases (NOES VNIIM collection)[3]

GSO number in the collection	RM index in the collection	Interval of permissible certified values of the **inferior calorific value**, $H_{inf}^{25/20}$, MJ/m[3a]	Permissible value relative to the extended uncertainty (U)[b] with a coverage factor k = 2, %
GSO 11,662–2020	NOES-VNIIM-H_2	10.030–10.260	0.3
GSO 11,663–2020	NOES-VNIIM-CH_4	33.350–33.580	
GSO 11,664–2020	NOES-VNIIM-C_2H_6	59.740–59.980	
GSO 11,665–2020	NOES-VNIIM-C_3H_8	86.230–86.470	

Notes [a] superscript (25/20) designates combustion reference conditions—temperature 25 °C (298,15 K) and pressure 101.325 kPa, and metering reference conditions—temperature 20 °C (293,15 K) and pressure 101.325 kPa
[b]corresponds to the permissible relative error at a confidence level (P = 0.95)

Table 5 Technical characteristics of initial materials of the rms of the inferior calorific value of gases (NOES VNIIM collection)

RM Index in the collection	Initial substance	Regulatory documents for the initial materials to comply with
NOES-VNIIM-H2	Hydrogen (H_2)	TU 2114–016-78,538,315–2008[4] or GOST R 51,673–2000[5]
NOES-VNIIM-CH4	Methane (CH_4)	TU 51 841–87[6]
NOES-VNIIM-C2H6	Ethane (C_2H_6)	TU 6–09-2454–85[7]
NOES-VNIIM-C3H8	Propane (C_3H_8)	TU 51–882-90[8]

The established metrological and technical requirements for the developed RMs are presented in Tables 4 and 5. Since the control of the metrological characteristics of calorimetric equipment is required at several points throughout the entire measurement range, pure gases with different values of the lowest volumetric combustion

[3] GSO 11,662–2020/GSO 11,665–2020 Standard Sample of the Lowest Volumetric Energy of Combustion of Gases (collection of NOES VNIIM—D.I. Mendeleyev All-Union Scientific Research Institute of Metrology). In: Federal Information Fund for Ensuring the Uniformity of Measurements. https://fgis.gost.ru/fundmetrology/registry/6/items/1384812 (In Russ.).

[4] TU 2114–016-78,538,315–2008 Highly Pure Hydrogen / NPK Nauka LLC. https://nauca.ru/content/hydrogen/HydrogenSpecification2114-016-78538315-2008.pdf (In Russ.).

[5] GOST R 51,673–2000 Gaseous Pure Hydrogen. Technical Specifications. Gosstandart Rossii, Moscow, 2000. (In Russ.).

[6] TU 51–841-87 Gaseous Methane / JSC Scientific Research and Design Institute of Urea and Organic Synthesis Products. http://www.1bm.ru/techdocs/kgs/tu/42/info/1351 (In Russ.).

[7] TU 6–09-2454–85 Isobutane for Chromatography and Spectroscopy of Chemically Pure Grade. https://www.standards.ru/document/3236550.aspx (In Russ.).

[8] TU 51–882-90 Liquefied Propane. https://www.standards.ru/document/3236550.aspx (In Russ.).

energy in the range of 10–90 MJ/m^3 were selected as the candidate substances in order to meet the need for metrological support. The target uncertainty of reference materials, taking into account the required ratio of the accuracy of the verification tool to the verified MI "1:3" was set at the level of 0.3%.

The choice of the technical characteristics of the initial materials was conditioned by the necessity to use commercially available substances.

Samples Under Study

When developing RMs, studies were carried out on the samples of pure gases, the technical characteristics of which are presented in Table 6.

When developing the RMs, the identification involved checking the accompanying documentation of the initial substances. The producer of initial materials (Monitoring Ltd., St. Petersburg) guarantees the compliance of pure gases with the requirements specified in Table 6, which is stated in the passports of initial materials. When developing the RMs, the gas analytical laboratory of VNIIM identified the main component and impurities for methane and propane. The results obtained showed that the molar fraction of the main component for the studied gases was not less than 99.99% mol., with the amount of each of the determined impurities (CO, CO_2, Ar + O_2, H_2, He, N_2) not exceeding 7 ppm.

The homogeneity and stability of the RMs were evaluated in accordance with Clause 7 and Clause 8 of ISO Guide 35:2017, taking into account the provisions for pure substances. In general, the chemical and physical properties of the studied pure substances suggest their high homogeneity and stability. According to Clause 7.1 of ISO Guide 35:2017, there is no need to assess between-unit heterogeneity, since each RM unit is assigned its own certified value. Some variability from within-unit heterogeneity inevitably manifests itself in standard deviations determined in characterization and stability studies; therefore, the contribution from within-unit heterogeneity was not evaluated separately, but was taken into account in aggregate with other contributions. In addition, the instructions for the RM provided the user with recommendations for storage and use of the RM to ensure its homogeneity. The

Table 6 Technical characteristics of the studied samples of the RMs of the lowest volumetric combustion energy of gases (NOES VNIIM collection)

Sample	Initial substance	Gas tank (material, volume)	Molar fraction of the main component, no less than, %
NOES-VNIIM-H2	Hydrogen (H_2)	Carbon steel, 10 l	99.95
NOES-VNIIM-CH4	Methane (CH_4)	Aluminum alloy, 8 l	
NOES-VNIIM-C2H6	Ethane (C_2H_6)	Carbon steel, 10 l	
NOES-VNIIM-C3H8	Propane (C_3H_8)	Aluminum alloy, 10 l	

Table 7 Results of measurements of volumetric combustion energy of hydrogen, methane, ethane, and propane samples under study

Sample (tank no.)	Date of the first series of measurements	Average value of inferior calorific value in the first series of the measurements $H_{inf,1}$, MJ/m^3	Date of the second series of the measurements	Average value of inferior calorific value in the second series of the measurements $H_{inf,2}$, MJ/m^3	Deviation between the average values of inferior calorific value in the first and second series, $\Delta(H_{inf})$, MJ/m^3 (%)
Hydrogen (1032)	November 1, 2016	10.052	October 11, 2019	10.051	0.001 (0.01)
Methane (27,890)	May 22, 2017	33.440	October 11, 2019	33.425	0.015 (0.04)
Ethane (15,722)	November 1, 2015	59.883	August 17, 2017	59.874	0.009 (0.01)
Propane (24,960)	October 17, 2019	86.410	February 15, 2021	86.365	0.045 (0.05)

RM stability was assessed on the basis of the data obtained during preliminary studies of similar initial materials according to Clause 8.2.3. The results of the stability studies are shown in Table 7.

The deviations obtained in the stability studies do not exceed 0.05%, which is less than one-third of the target uncertainty of the RM inferior calorific value (0.3%). Therefore, the contribution from instability to the total uncertainty can be considered negligible.

Method for Determining the Certified Value

The standard gas calorimeters-comparators USVG and USNG, included in GET 16–2018 in the course of its improvement completed in 2018, were used to determine the certified value of the combustion energy of the studied samples. The expanded uncertainty of the measurement results in the range of 10–90 MJ/m^3, obtained with these calorimeters, is estimated at a level not exceeding 0.3%.

The standard gas calorimeters are designed to transfer a unit of volumetric energy of combustion to gases and to gas mixture in the range of 3–35 MJ/m^3 (USNG calorimeter) and 25–90 MJ/m^3 (USVG calorimeter). The calorimeters implement the direct method of measuring the volumetric heat of combustion, connected by the calibration dependence with the measured combustion time of a single gas portion.

Figure 1 shows the principle of operation of the calorimeters. Gas combustion is performed in the burner of the measuring cell (1) located inside the heating unit (2). The gas is fed from the closed tank (3) through the gas line (4) to the measuring cell

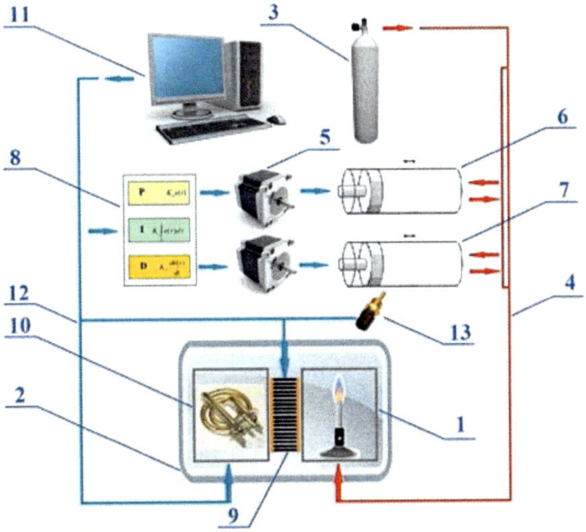

Fig. 1 The principle of operation of the reference gas calorimeters-comparators USVG and USNG included in the GET 16–2018. 1—burner of the measuring cell; 2—heating unit; 3—tank; 4—gas line; 5—stepper motors; 6 and 7—working cylinders; 8—PID controller; 9—differential module; 10—comparative cell; 11—personal computer; 12—system of auxiliary electronic units and electrical cables; 13—sensors

by a dosing device with an adjustable feed rate. The gas supply rate is controlled by changing the frequency of pulses supplied to the stepper motors (5), which with the help of worm-screw pairs drive the pistons in the working cylinders (6 and 7) into translational movements. The frequency control is performed by a PID controller (8), which maintains zero signal on the set of differential modules (9), such that the thermal power, released in the measuring cell due to the test gas combustion, is constant and equal to the power released in the comparative cell (10).

The power comparison method is based on the compensation method implemented on the basis of a differential thermal circuit with diathermic coupling through semiconductor converters, which are zero-organs between the cells—a measuring cell, in which the gas combustion process is carried out, and a comparative cell, in which an electric heater is located, which releases constant power.

The operation of the calorimeters is controlled by means of a personal computer (11) of industrial design with a system of auxiliary electronic units and electrical cables (12) forming control, regulation, power supply, and coordination units.

The calorimeter measurement mode is continuous-cyclic. Continuity is ensured by alternating operation of two piston pumps. Each measurement cycle is associated with extruding a portion of gas by one of the pumps. To bring the measurement results of the volumetric LCV to standard conditions, the metering device has a system of sensors (13), which provide the formation of signals that are sent to the control and regulation system for processing.

Statistical results were obtained by processing the array of measured data (single values of the volumetric LCV, issued by the calorimeter at the end of each measuring cycle), with the extraction of data for each of the studied gases from the general array with the measurement results.

Estimated Uncertainty of the Certified Value

Since the input values are independent, in accordance with ISO Guide 35:2017, the total standard uncertainty of the certified value was calculated with the following equation [19]:

$$u_{CRM} = \sqrt{u_{char}^2 + u_{hom}^2 + u_{lts}^2}, \tag{2}$$

where

u_{char}—uncertainty of the certified value obtained for the unit (characterization);

u_{hom}—uncertainty associated with the heterogeneity of the material in a single package (homogenity);

u_{lts}—uncertainty associated with material instability (stability).

Since the volumetric combustion energy is an additive value and directly depends on the component composition of the gas, the homogeneity and stability of RMs were confirmed by the positive results of experimental studies of the stability of RMs of the component composition (reference standards), which were manufactured using similar initial materials. The components of heterogeneity and instability were evaluated as negligible and were excluded from the estimation of total uncertainty.

The characterization of uncertainty u_{char} was estimated in accordance with the uncertainty budget of the USVG and USNG calorimeters:

$$u_{char} = u_c(H), \tag{3}$$

$$u_c(H) = \sqrt{u_b(H)^2 + u_a(H)^2}, \tag{4}$$

where

$u_b(H)$—the relative standard uncertainty of type B measurements of the volumetric energy of gas combustion;

$u_a(H)$—the relative standard uncertainty of type A measurements of the volumetric energy of gas combustion, calculated by the formula:

$$u_A(H) = \sqrt{\frac{\sum_{i-1}^{n} (H_i - \overline{H})^2}{n-1}}, \tag{5}$$

where

H_i—value of the *i*th measurement result of the LCV, MJ/m^3;

$\overline{H}$—average value of measuring the LCV, MJ/m^3;

n—number of single measurement results.

The uncertainty $u_b(H)$ when transferring a unit of combustion energy using the USVG and USNG calorimeters was established earlier in the course of their studies as part of the improvement of GET 16–2018 and in the range of 10–90 MJ/m does not exceed $8 \cdot 10^{-4}$.

The expanded uncertainty was calculated by the formula

$$U_{CRM} = k \cdot u_{CRM}, \quad (6)$$

where the coverage factor is $k = 2$ with a confidence level of $P = 0.95$.

Results and Discussion

In the course of experimental studies, a certified value of inferior calorific value, traceable to the State Primary Standard, was obtained for the 4 samples under study, and its uncertainty was evaluated. The results of experimental studies are shown in Table 8.

Figures 2, 3, 4 and 5 show the results of measurements of the inferior calorific value of the studied samples. In order to obtain a sufficient amount of initial data to determine the desired value and estimate the budget of measurement uncertainty, each sample was burned for 24 h.

The stability and absence of a drift of the combustion energy of the studied samples additionally testify to the correctness of the choice of initial materials with the aim to develop on their basis the metrological means for verification, calibration, and testing of gas calorimetric equipment used in industry.

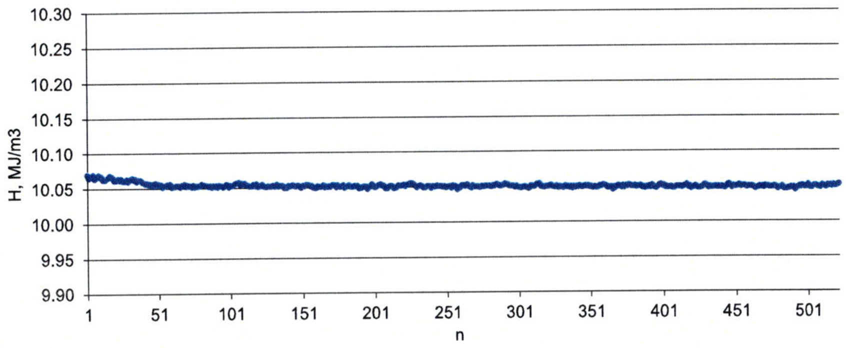

Fig. 2 Measurement of inferior calorific value when characterizing a hydrogen sample (USNG calorimeter-comparator)

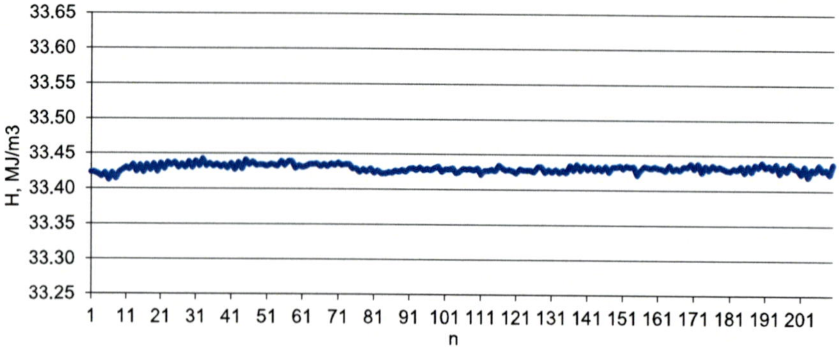

Fig. 3 Measurement of inferior calorific value when characterizing a methane sample (USVG calorimeter-comparator)

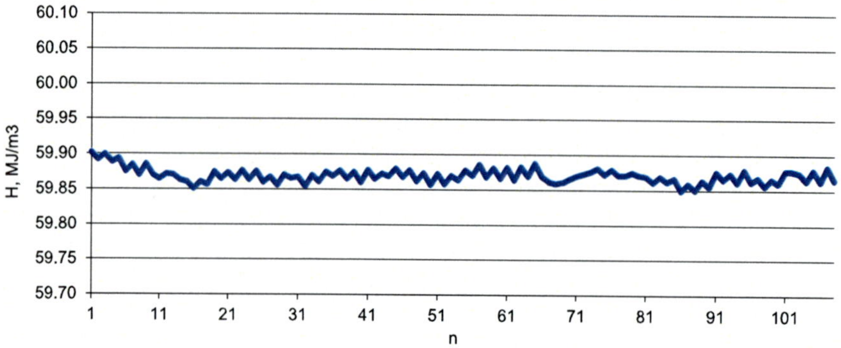

Fig. 4 Measurement of inferior calorific value when characterizing an ethane sample (USVG calorimeter-comparator)

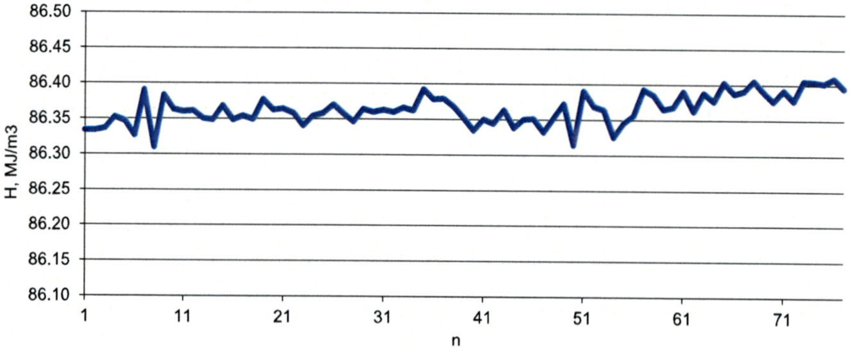

Fig. 5 Measurement of inferior calorific value when characterizing a propane sample (USVG calorimeter-comparator)

Table 8 Measurements of volumetric combustion energy of hydrogen, methane, ethane, and propane samples under study

Sample (calorimeter)	Measured value of inferior calorific value H_{inf}, MJ/m^3	SD of the measurement result, s_r, MJ/m^3 (in 24 h, n—number of single measurement results)	Standard characterization uncertainty, u_{char}, MJ/m^3	Relative expanded uncertainty U_{crm} at coverage factor $k = 2$, MJ/m^3 (%)
Hydrogen (USNG)	10.053	0.004 ($n = 523$)	0.009	0.018 (0.18)
Methane (USVG)	33.431	0.005 ($n = 210$)	0.027	0.054 (0.16)
Ethane (USVG)	59.870	0.010 ($n = 108$)	0.079	0.158 (0.26)
Propane (USVG)	86.365	0.022 ($n = 77$)	0.115	0.230 (0.27)

The additional characteristics of the reference materials—relative density and the Wobbe index—can be further calculated using the equations of state given in the GSSSD standards[9] [20–22] (Table 8).

Conclusion

In this article, we considered the possibility of developing a certified RM for inferior calorific value based on pure gases—hydrogen, methane, ethane, and propane. On the basis of our studies, we developed and certified a reference material of the approved type of the inferior calorific value of gases (collected at NOES VNIIM) GSO 11,662–2020/GSO 11,665–2020. The RM set is designed to establish and control the metrological characteristics of gas calorimeters and Wobbe index analyzers. The following certified inferior calorific values were established for the studied samples of pure gases: hydrogen (10.053 ± 0.018) MJ/m^3, methane (33.431 ± 0.054) MJ/m^3, ethane (59.870 ± 0.158) MJ/m^3, and propane (86.365 ± 0.230) MJ/m^3. The certified values were obtained using a direct calorimetric method, traceable to the State Primary Standard GET 16, and meet the requirements for metrological characteristics given in the description of the RM type and in Table 4. In the future, the applied method will expand the metrological support of this measurement area by expanding the range of RMs of gas mixtures—imitators of combustible gases. The analysis also showed the need for RMs of the Wobbe index, the development of which requires

[9] GSSSD—Gosudarstvennaya sluzhba standartnykh spravochnykh dannykh o fizicheskikh konstantakh i svoystvakh veshchestv i materialov [State Service for Standard Reference Data on the Physical Constants and Properties of Substances and Materials].

further research, in particular, the choice of a method for determining the value of the relative density of gases.

Acknowledgements The authors are grateful to the staff of the Laboratory of Research in the Field of Gas Analytical Measurements of the D. I. Mendeleev Institute for Metrology (VNIIM), especially to Anna Kolobova and Natalya Uvarova, for their expertise and assistance in preparing the documentation during the approval of the reference material.

Author Contributions K. A. Mishina: literature review; data analysis and discussion; initial draft preparation.

E. N. Korchagina: research supervision, text revision and editing.

I. V. Kazartsev: data collection and analysis; text revision and editing.

Conflict of Interest The article was prepared on the basis of a report presented at the IV International Scientific Conference "Reference Materials in Measurement and Technology" (St. Petersburg, December 1–3, 2020). The article was admitted for publication after the abstract was revised, the article was formalized and the review procedure was carried out.

The version in the Russian language is published in the journal "Measurement Standards. Reference Materials" 2021;17(2):19–32. https://doi.org/10.20915/2687-0886-2021-17-2-3519-32.

References

1. Global gas report 2018. http://www.snam.it/export/sites/snam-rp/repository/file/gas_naturale/global-gas-report/global_gas_report_2018.pdf
2. Malek L, Hulteberg C (2016) Measuring and ensuring the gas quality of the Swedish gas grid. Energiforsk, p 38. https://portal.research.lu.se/portal/en/publications/measuring-and-ensuring-the-gas-quality-of-the-swedish-gas-grid(e8249698-1599-444f-992f-d6fb19a63b48).html
3. GOST 31369–2008 Gaz prirodnyj. Vychislenie teploty sgoranija, plotnosti, otnositel'noj plotnosti i chisla Vobbe na osnove komponentnogo sostava. Standartinform, Moscow, p 54 (In Russ.)
4. ISO 15971:2008 Natural gas—Measurement of properties—Calorific value and Wobbe index. https://www.iso.org/standard/44867.html
5. Dörr H, Koturbash T, Kutecherov V (2019) Review of impacts of gas qualities with regard to quality determination and energy metering of natural gas. Meas Sci Technol 30(4):022001. https://doi.org/10.1088/1361-6501/aaeef4
6. OIML R 140: 2007 (E) Measuring systems for gaseous fuel. https://www.oiml.org/en/files/pdf_r/r140-e07.pdf
7. Villermaux C, Zarea M, Haloua F, Hay B, Filtz J-R (2006) Measurement of gas calorific value: a new frontier to be reached with an optimised reference gas calorimeter. In: 23rd World gas conference. Amsterdam. http://members.igu.org/html/wgc2006/pdf/paper/add12646.pdf
8. Jaeschke M, Schmücker A, Pramann A, Ulbig P (2007) GERG project: development and setup of a new combustion reference calorimeter for natural gases. Int J Thermophys 28:220–244. https://doi.org/10.1007/s10765-007-0167-1
9. ISO 6976:2016 Natural gas — Calculation of calorific values, density, relative density and Wobbe indices from composition. https://www.iso.org/ru/standard/55842.html
10. Haloua F, Foulon E, Allard A, Hay B, Filtaz J-R (2015) Traceable measurement and uncertainty analysis of the gross calorific value of methane determined by isoperibolic Calorimetry. Metrologia 52(6):741–755. https://doi.org/10.1088/0026-1394/52/6/741

11. Rauch J, Haloua F (2018) Measurements of the Calorific value of Methane with the New GERG reference Calorimeters. J Phy Conf Series 1065(20). https://doi.org/10.1088/1742-6596/1065/20/202007
12. Alexandrov YuI, Chunovkina AG, Korchagina EN (2002) Revised value of the heat of combustion for high purity methane. In: Proceedings conference and exhibition on natural gas quality. NPL, Loughboroug, UK, 26–28 November, p 7
13. Perez-Sanz FJ, Sarge SM, van der Veen A, Culleton L, Beaumont O, Haloua F (2019) First experimental comparison of calorific value measurements of real biogas with reference and field calorimeters subjected to different standard methods. Int J Therm Sci 135:72–82. https://doi.org/10.1016/j.ijthermalsci.2018.06.034
14. Kacur J, Kostur K (2015) Indirect measurement of syngas calorific value. In: Proceedings of the 2015 16th international Carpathian control conference (ICCC). Szilvasvarad, Hungary, 27–30 May, 2015. pp 229–234. https://doi.org/10.1109/CarpathianCC.2015.7145079
15. Rauch J, Haloua F (2018) Calorific value of biomethane: comparative measurements using reference gas calorimeters value. J Phy Conf Series 1065(20). https://doi.org/10.1088/1742-6596/1065/20/202007
16. Tsochatzidis NA, Karantanas E (2012) Assessment of calorific value at a gas transmission network. J Nat Gas Sci Eng 9:45–50. https://doi.org/10.1016/j.jngse.2012.05.009
17. Haloua F, Foulon E, El-Harti E, Sarge SM, Rauch J, Neagu M, Brown AS, Tuma D (2016) Comparison of traceable methods for determining the calorific value of non-conventional fuel gases. Int J Therm Sci 100:438–447. https://doi.org/10.1016/j.ijthermalsci.2015.10.020
18. Ulbig P, Hoburg D (2002) Determination of the calorific value of natural gas by different methods. Thermochim Acta 382(1–2):37–35. https://doi.org/10.1016/S0040-6031(01)00732-8
19. ISO Guide 35:2017 Reference materials—Guidance for characterization and assessment of homogeneity and stability. https://www.iso.org/standard/60281.html
20. GSSSD 195–01 (2008) Metan zhidkij i gazoobraznyj. Termodinamicheskie svojstva, kojefficienty dinamicheskoj vjazkosti i teploprovodnosti pri temperaturah 91...700 K i davlenijah 0,1...100 MPa. Standartinform, Moscow, p 31 (In Russ.)
21. GSSSD 196–01 (2008) Tablicy standartnyh spravochnyh dannyh. Jetan zhidkij i gazoobraznyj. Termodinamicheskie svojstva, kojefficienty dinamicheskoj vjazkosti i teploprovodnosti pri temperaturah 91...625 K i davlenijah 0,1...70 MPa. Standartinform, Moscow, p 36 (In Russ.)
22. GSSSD 197–01 (2008) Tablicy standartnyh spravochnyh dannyh. Propan zhidkij i gazoobraznyj. Termodinamicheskie svojstva, kojefficienty dinamicheskoj vjazkosti i teploprovodnosti pri temperaturah 86...700 K i davlenijah 0,1...100 MPa. Standartinform, Moscow, p 38 (In Russ.)

Mishina Karina A. Engineer, D. I. Mendeleyev Institute for Metrology (VNIIM).

Korchagina Elena N. Cand. Sci., Head of the Laboratory for Measurement Standards and Scientific Research in the field of combustion calorimetry and high-purity organic substances for metrological purposes, D. I. Mendeleyev Institute for Metrology (VNIIM).

Kazartsev Iaroslav V. Researcher, D. I. Mendeleyev Institute for Metrology (VNIIM).

Selection of Materials for Novel Reference Materials of Fluid Viscosity Certified in the Interval of Permissible Temperature Values from Minus 40 °C to Minus 5 °C

Anastasia A. Neklyudova and Aleksey A. Demyanov

Abstract We present a research study aimed at selecting materials for the production of novel reference materials (RMs) intended for verification, calibration, and testing of instruments for measuring the viscosity of liquids, as well as for certification of methods for measuring the viscosity of liquid media at temperatures below 0 °C. The developed RMs are certified in the range of permissible temperature values from minus 40 °C to minus 5 °C, which allows a wide range of modern instruments for measuring the viscosity of liquids, such as glass capillary viscometers, rotational viscometers, rheometers, etc., to be provided with metrological supervision.

Keywords Approved reference materials · Dynamic viscosity · Temperature · Traceability · Ensuring the uniformity of measurements

Introduction

The oil industry plays an important role in the economy of any country, including the Russian Federation. Currently, 32 large oil refineries [1] and numerous small oil refineries operate in Russia with the processing volumes of more than one million tons per year. In terms of total capacity, the Russian oil refining industry ranks third in the world, behind the United States and China [2].

The five largest oil refining companies in Russia include Gazprom Neft PJSC, Surgutneftegaz OJSC, Rosneft PJSC, Oil Company Lukoil PJSC, and ANK Bashneft PJSC [3]. In accordance with the "Development of Energy" National Program [4] for the period from 2013 to 2024, it is planned to increase the depth of oil refining and the output of fuel that meets technical regulations, as well as to perform the construction, modernization and reconstruction of oil refineries.

A. A. Neklyudova (✉) · A. A. Demyanov
D. I. Mendeleyev Institute for Metrology (VNIIM), Moskovsky pr., 19, Saint Petersburg, St.-Petersburg 190005, Russia
e-mail: A.A.Tsurko@vniim.ru

A. A. Demyanov
e-mail: A.A.Demyanov@vniim.ru

S. V. Medvedevskikh et al. (eds.), *Reference Materials in Measurement and Technology*, https://doi.org/10.1007/978-3-031-06285-8_11

All these measures should ensure not only an increase in the amount of oil products and a decrease in ineffective processing of raw materials, but also improve the quality of products. The main products of oil refineries include gasolines, kerosene, aviation, and rocket fuels, fuel oils, diesel fuels, oils, lubricants, bitumen, petroleum coke, etc.

When assessing the quality of petroleum products, along with such parameters as density and fractional composition, viscosity coefficient is widely used. In turn, the correctness of the decision taken during the technological control of the quality of manufactured products depends on the accuracy of determining this coefficient. The employed measuring instruments (MIs), measurement techniques (MTs), and the qualifications of the operator have the greatest influence on the accuracy of measurements.

Today, there is a huge amount of viscosity MIs, which are used in the oil refining industry, related to the field of state regulation of ensuring the uniformity of measurements. Thus, in accordance with No. 102-FZ Federal Law "On Ensuring the Uniformity of Measurements", those MIs must undergo primary and periodic verification [5].

The verification of viscosity MIs, in accordance with the document "State system for ensuring the uniformity of measurements. State verification scheme for measuring instruments for viscosity of liquids", approved by the order No. 2622 of the Federal Agency for Technical Regulation and Metrology dated 05.11.2019 [6], should be carried out using working reference materials (RMs) of the second category, which are RMs of fluid viscosity, or by the method of direct comparison using calibration fluids (comparators).

The need for a large nomenclature of RMs of fluid viscosity is determined by the variety of MI types in the field of viscometry, since there is a problem of metrological support for MIs determining the viscosity of liquids in the temperature range from minus 40 °C to minus 5 °C. Those measuring instruments, as a rule, are used to control the quality of engine oils and lubricants, which implies the verification of these MIs using RMs.

Thus, the development of stable and homogeneous RMs of fluid viscosity, certified in the range of permissible temperatures from minus 40 °C to minus 5 °C, is an urgent research task.

Materials and Methods

Development and Study of RMs

The development of RMs of fluid viscosity, certified in the range of permissible temperatures from minus 40 °C to minus 5 °C, was carried out within the framework of improving the GET 17–96[1] national primary standard.

RM development included six stages:

(1) collection and analysis of the initial information; preparation of a specification of requirements for the development of an RM, its metrological examination and approval;
(2) selection of starting materials, methods of RM preparation, methods of describing their characteristics, labelling, storage and transportation; development of a metrological traceability scheme;
(3) preparatory work: calibration (verification) and equipment suitability check; ensuring the necessary conditions in the premises for the RM production, development and assessment of the suitability (validation, certification) of measurement procedures; staff training;
(4) production of RM prototypes (batch) and their measurement to establish (assign) the values of RM characteristics;
(5) assessment of the RM homogeneity and stability, as well as the uncertainty of the RM certified values;
(6) preparation of documents for the developed RMs.

When selecting starting materials for the development of an RM of fluid viscosity, we conducted a study of the homogeneity of the Mobil 1 10 W-60 (Mobil passenger-vehicle-lube, Russia) engine oil, as a negative example of the use of such materials in the production of RMs of fluid viscosity.

The material homogeneity of a Mobil 1 10 W-60 engine oil sample was tested using a capillary measurement method. A representative sample of the engine oil material was taken from the top half of the vial as well as from the bottom half of the vial as shown in Fig. 1.

The results of measuring the kinematic viscosity of the Mobil 1 10 W-60 engine oil at minus 30 °C using the capillary measurement method are given in Table 1.

Table 1 shows that the relative sample standard deviation of the arithmetic mean obtained during this experiment reaches 0.4%. This is a rather large value, since this parameter usually does not exceed 0.1% [6] when measuring the viscosity of homogeneous liquids using the capillary measurement method and using RMs of the 1st category.

The repeatability of the results obtained for the kinematic viscosity of engine oil between sample No. 1 and No. 2, taken from one sample unit, is 1.4%. This

[1] GET 17–96 National primary standard of the unit of kinematic viscosity of a liquid. In: Federal Informational Fund on Maintaining the Unity of Measurements. https://fgis.gost.ru/fundmetrology/registry/12/items/947620.

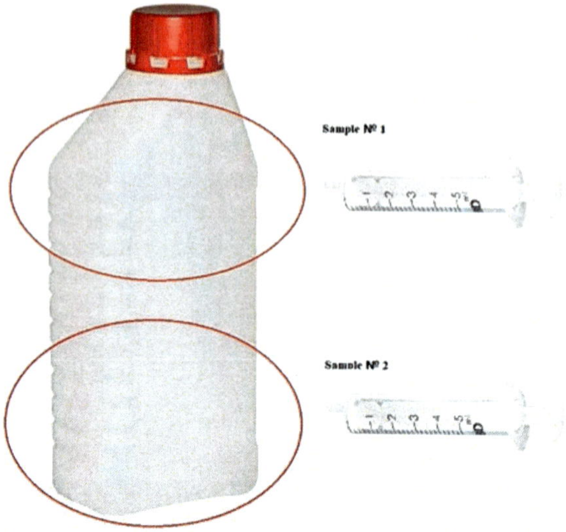

Fig. 1 The sampling scheme for the Mobil 1 10 W-60 engine oil

Table 1 The kinematic viscosity of the Mobil 1 10 W-60 engine oil

Parameters	Sample number	
	1	2
Average measured value of the flowing time, s	157.62	159.78
Relative sample standard deviation of the arithmetic mean, %	0.2	0.4
Kinematic viscosity of engine oil at $T = -30$ °C	8262.8	8029.5
Convergence, %	1.4	

also indicates that, although this material meets the requirements for the quality of engine oils of that class, it cannot be used as a RM material due to its significant heterogeneity [7]. Other experimental results also testify to the inappropriateness of using engine oil as a material for an RM.

The dynamic viscosity of a sample produced on the basis of the TNK Revolux D2 10 W-40 engine oil (Rosneft, Russia) by the rotary method at a temperature $T = (-25.00 \pm 0.05)$ °C on a Physica MCR 301 Rheometer (Anton Paar GmbH, Germany) was studied. The experiments were aimed at establishing the repeatability of the measured values of the dynamic viscosity of the TNK 10 W-40 engine oil at a shear rate of 38.6 s^{-1} and five loadings of the sample material (five series of measurements with 15 observations) into the measuring system by one operator using the same measuring instrument. The results are presented in Table 2 and Fig. 2.

The research also found that this engine oil, containing a package of additives, is a non-Newtonian fluid [8], i.e. the dependence of the dynamic viscosity of the TNK 10 W-40 engine oil on shear rate is observed.

Figure 3 shows the dependence of the dynamic viscosity of the TNK 10 W-40 on the shear rate in the range from 0 min^{-1} to 60 min^{-1}.

Table 2 The dynamic viscosity of a sample based on the TNK 10 W-40 engine oil at a temperature $T = (-25.00 \pm 0.05)$ °C

Measurement series number	Average measured value of dynamic viscosity, mPa·s	Rotational speed of the measuring rotor, min^{-1}	Shear rate, s^{-1}	Relative repeatability of the average measured values of dynamic viscosity, %
1	11,170	30.0	38.6	–
2	11,240			+ 0.6
3	11,370			+ 1.8
4	11,350			+ 1.6
5	11,300			+ 1.2

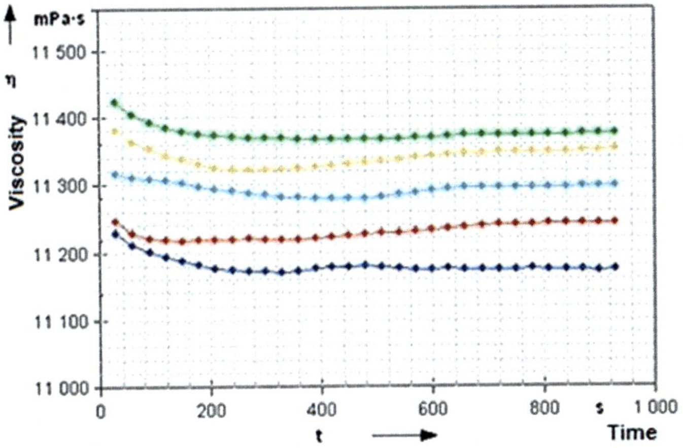

Fig. 2 The repeatability of the measured dynamic viscosity of a sample produced on the basis of the TNK engine oil at a constant shear rate

According to quality certificates, ready-made for use engine oils contain mechanical impurities, whose mass fraction can reach 0.015%. This can also affect the quality of an RM produced from such a material.

Thus, it was concluded that the use of ready-made engine oils is not an acceptable option for the production of RMs of fluid viscosity, since the additives contained therein negatively affect the stability and homogeneity of such materials. In addition, such oils are non-Newtonian fluids, which entails the need to control the shear rate in the experiment.

In order to identify the most stable and homogeneous materials suitable for the production of viscosity RMs, we analyzed modern base oils, free from additives and mechanical impurities. For the research, synthetic and mineral motor oils, as well as mixtures of the latter, were selected.

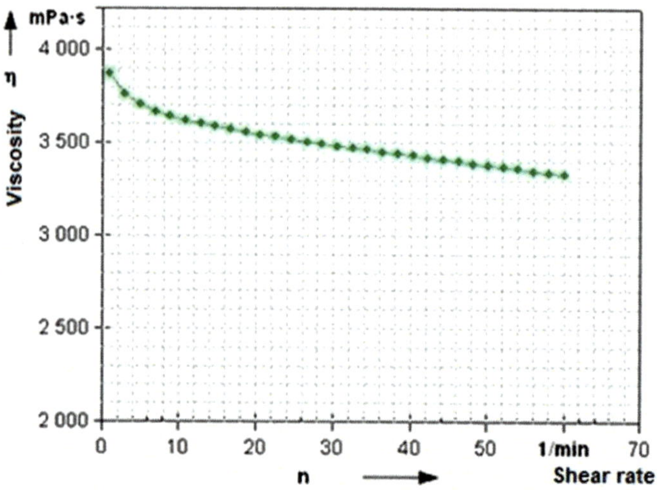

Fig. 3 The dependence of the dynamic viscosity of a sample based on the TNK 10 W-40 engine oil on shear rate

For the production of low-temperature RMs, base synthetic motor oils of group 4— polyalphaolefins (PAO)—were selected.

The world's leading manufacturers of polyalphaolefins are: Idemitsu Kosan Co.—Lenealene, Japan; Exxon Mobil—SpectraSyn, USA and France; INEOS—Durasyn, USA and Belgium; Chemtura—Synton, Canada; ChevronPhillips—SynFluid, USA; Tatneft—Nizhnekamskneftekhim, Russia.

To confirm the suitability of the selected materials as viscosity RMs, experiments were carried out to establish the repeatability of the measured values of dynamic viscosity using the rotary method and a Physica MCR 301 rheometer, similar to the experiments with engine oil. The speed of rotation of the measuring rotor was also 30 min^{-1}, the number of the loads of the PAO-170 material was five.

The results of measuring the dynamic viscosity of a sample based on the PAO-170 at a temperature $T = (-25.00 \pm 0.05)$ °C are shown in Table 3 and Fig. 4.

In the course of the research, we also found that this base oil, containing no additives, is a Newtonian fluid, i.e. no dependence of the dynamic viscosity on the shear rate was observed. Experiments were carried out in the range of rotational speeds of the measuring rotor from 1 min^{-1} to 60 min^{-1}.

Figure 5 shows the dependences of the dynamic viscosity of the studied samples based on motor oil and polyalphaolefin. The experimental temperatures were selected such that the nominal values of the dynamic viscosity of the samples ranged from 3000 mPa·s to 5000 mPa·s, i.e. were comparable.

When carrying out the measurements of the kinematic viscosity of synthetic base and engine oils, structural changes in the material of the engine oil were visually established. Thus, when poured into a capillary glass viscometer and placed in a thermostatic bath with the temperature set to minus 30 °C, the oil from a yellow

Table 3 The dynamic viscosity of a sample based on the PAO-170 at a temperature $T = (-25.00 \pm 0.05)$ °C

Measurement series number	Average measured value of dynamic viscosity, mPa·s	Rotational speed of the measuring rotor, min^{-1}	Shear rate, s^{-1}	Relative repeatability of the average measured values of dynamic viscosity, %
1	4254.0	30.0	38.6	–
2	4261.4			+0.2
3	4254.8			+0.02
4	4259.5			+0.1
5	4271.9			+0.4

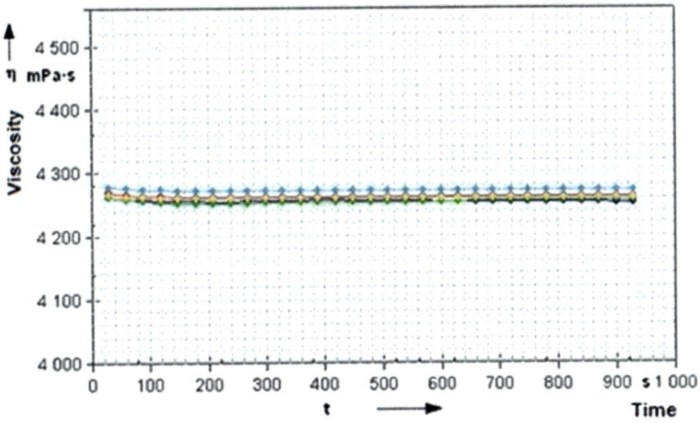

Fig. 4 The repeatability of the measured dynamic viscosity of a sample produced on the basis of the PAO-170 at a constant shear rate

transparent liquid turned into a white-yellow cloudy substance. This effect was not observed when polyalphaolefin oils were placed under the same conditions.

Figures 6 and 7 present photographs of motor and polyalphaolefin oils placed in a capillary glass viscometer in a thermostatic bath.

Results and Discussion

We found that, although the Mobil 1 10 W-60 and TNK Revolux D2 10 W-40 engine oils are ready-to-use, i.e. containing a complex of additives and meeting the requirements for oils used in internal combustion engines, these oils cannot be used for preparing RMs due to their high instability and heterogeneity. The

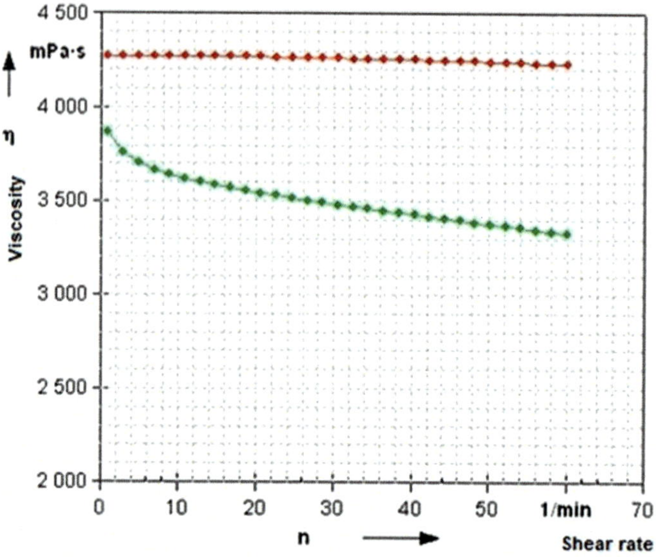

Fig. 5 The dependence of the dynamic viscosity of samples based on motor oil and polyalphaolefin on shear rate

Fig. 6 Engine oils placed in a capillary glass viscometer in a thermostatic bath set at minus 30 °C

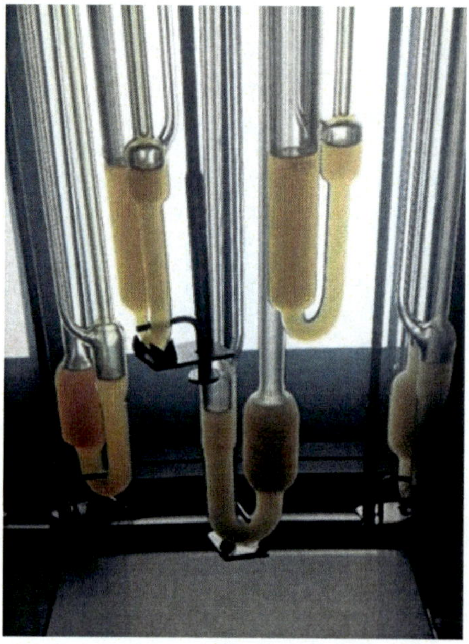

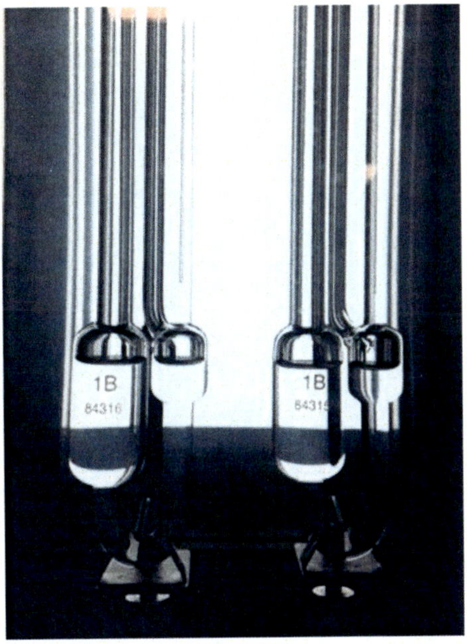

Fig. 7 Polyalphaolefin oil placed in a capillary glass viscometer in a thermostatic bath set at minus 30 °C

base motor oils of group 4 (polyalphaolefins from Idemitsu Kosan Co. —Lenealene, Exxon Mobil—SpectraSyn, INEOS—Durasyn, Chemtura—Synton, Chevron-Phillips—SynFluid, and Tatneft—Nizhnekamskneftekhim), containing no additives and mechanical impurities, are not suitable for use in engines in their pure form. However, the high stability and homogeneity of these oils render them suitable for producing RMs of fluid viscosity.

We developed three CRMs: GSO 11,516–2020 REV-VNIIM-50, GSO 11,517–2020 REV-VNIIM-100, and GSO 11,518–2020 REV-VNIIM-200. The metrological characteristics of the developed CRMs, certified in the range of permissible temperature values from minus 40 °C to minus 5 °C, are presented in Tables 4, 5 and 6.

Conclusion

The developed CRMs were tested in more than 20 metrological projects, such as verification, calibration, and testing for the purpose of approving MI types. These CRMs solve not only the problem of the lack of verification tools for viscosity measuring instruments in the temperature range from minus 40 °C to minus 5 °C, but have also initiated comparisons of national standards for the unit of kinematic viscosity of a liquid. The proposal to launch international comparisons was announced by the representatives of Mendeleyev Institute for Metrology (VNIIM) at the meeting of the CCM.WGDV working group held on the 13th of May, 2019 in Sèvres (France).

Table 4 Metrological characteristics of the GSO (CRM) 11,516–2020 REV-VNIIM-50

Certified characteristic	The interval of permissible certified values measured at temperature, °C							
	– 40.00 ± 0.02	– 35.00 ± 0.02	– 30.00 ± 0.02	– 25.00± 0.02	– 20.00 ± 0.01	– 15.00 ± 0.01	– 10.00 ± 0.01	– 5.00± 0.01
Kinematic viscosity, ν, mm^2/s	6870–9300	3715–5035	2195–2975	1155–1565	865–1200	574–777	392–530	274–370
Dynamic viscosity, η, mPa·s	5930–8020	3195–4325	1875–2545	985–1335	735–1000	485–657	330–450	230–315
Permissible values of the relative expanded uncertainty of the certified values of dynamic and kinematic viscosity $(U_R)^a$, at $k = 2$,%	0.4				0.3			

[a] –corresponds to the boundaries of the relative error ± δ, % with a confidence level P = 0.95

Table 5 Metrological characteristics of the GSO (CRM) 11,517–2020 REV-VNIIM-100

Attested characteristic	Interval of permissible certified values measured at temperature, °C							
	– 40.00±	– 35.00±	– 30.00±	– 25.00±	– 20.00±	– 15.00±	– 10.00±	– 5.00±
	0.02	0.02	0.02	0.02	0.01	0.01	0.01	0.01
Kinematic viscosity, ν, mm^2/s	14930 – 20210	7515 – 10250	4270 – 5780	2560 – 3470	1590 – 2160	870 – 1180	685 – 925	470 – 640
Dynamic viscosity, η, mPa·s	12930 – 17500	6490 – 8780	3675 – 4975	2195 – 2975	1360 – 1840	740 – 1010	580 – 790	395 – 545
Permissible values of the relative expanded uncertainty of the certified values of dynamic and kinematic viscosity $(U_R)^a$, at $k = 2$,%	0.4				0.3			

[a] –corresponds to the boundaries of the relative error ± δ, % with a confidence level P = 0.95

Table 6 Metrological characteristics of the GSO (CRM) 11,518–2020 REV-VNIIM-200

Attested characteristic	Interval of permissible certified values measured at temperature, °C							
	– 40.00±	– 35.00±	– 30.00±	– 25.00±	– 20.00±	– 15.00±	– 10.00±	– 5.00±
	0.02	0.02	0.02	0.02	0.01	0.01	0.01	0.01
Kinematic viscosity, ν, mm^2/s	29350 – 39710	14380 – 19460	7835 – 10605	4510 – 6105	2720 – 3680	1450 – 1970	1105 – 1495	625 – 850
Dynamic viscosity, η, mPa·s	25500 – 34500	12445 – 16835	6760 – 9140	3875 – 5242	2330 – 3150	1240 – 1680	935 – 1270	530 – 720
Permissible values of the relative expanded uncertainty of the certified values of dynamic and kinematic viscosity $(U_R)^a$, at $k = 2$,%	0.4				0.3			

[a] –corresponds to the boundaries of the relative error ± δ, %, with a confidence level P = 0.95

Author Contributions All authors have made an equivalent contribution to the preparation of the publication.

Conflict of Interest The article was prepared on the basis of a report presented at the IV International Scientific Conference "Reference Materials in Measurement and Technology" (St. Petersburg, December 1–3, 2020). The article was admitted for publication after the abstract was revised, the article was formalized and the review procedure was carried out.

The version in the Russian language is published in the journal "Measurement Standards. Reference Materials" 2021;17(3):63–71. https://doi.org/10.20915/2687-0886-2021-17-3-63-71.

References

1. Ryabov VA (2019) On crisis problems in the oil refining and petrochemical industry. Siberian Oil and Gas 1(34). http://sib-ngs.ru/journals/article/978 (In Russ)
2. Refining of oil and gas condensate. https://minenergo.gov.ru/node/1209 (In Russ)
3. (2017) The largest oil refineries in Russia. *Pronedra* 25 Apr. https://pronedra.ru/oil/2017/04/25/krupneyshie-npz-rossii/
4. State program "Development of energy": open ministry. https://minenergo.gov.ru/node/323
5. Federal Law (2008) № 102-FZ of 26/06/2008 "On Ensuring the Uniformity of Measurements" (as amended and supplemented). https://normativ.kontur.ru/document?moduleId=1&documentId=398044 (In Russ)
6. On the approval of the State verification scheme for measuring instruments for the viscosity of liquids: Order No. 2622 dated November 05, 2019. https://www.rst.gov.ru/portal/gost/home/activity/documents/orders#/order/126301 (In Russ)
7. GOST ISO Guide 34–2014 (2019) General requirements for the competence of reference material producers. Standartinform, Moscow, 2019, p 39 (in Russ)
8. McKelvey DM (1965) Processing of polymers, trans. from English. Chemistry, Moscow, p 444

Anastasia A. Neklyudova Cand. Sci., Deputy Head of Density and Viscosity Laboratory, D. I. Mendeleyev Institute for Metrology (VNIIM).

Aleksey A. Demyanov Head of Density and Viscosity Laboratory, D. I. Mendeleyev Institute for Metrology (VNIIM).

Development of a Metrological Support Complex for the Food Industry

Anna S. Sergeeva, Natalia L. Vostrikova, and Maria Yu. Medvedevskikh

Abstract The contribution of experts from the Laboratory for Metrological Support of Moisture Measurement and Reference Materials to improving the metrological support system for the food industry is considered. The article summarizes over 40 certified reference materials (CRMs) for the composition of food products, raw materials, and additives developed from 2008 to 2020. The development of each new CRM included the following steps: material preparation; stability and homogeneity study thereof; establishment of the certified value using state primary (GET 173) and secondary (GVET 176–1) measurement standards; primary reference measurement procedures; an interlaboratory experiment for checking the applicability of CRMs. The creation of CRMs for the composition of grain and powdered milk products accompanied GET 173 development in 2008. The approval of GVET 176–1 in 2010 allowed the certified value of nitrogen (protein) mass fraction to be added to CRMs. The successful participation of UNIIM in comparisons resulted in the publication of 6 CMC (calibration and measurement capabilities) lines for the measurement of nitrogen mass fraction in glycine, milk powder, grain, egg powder, porridge, and feed in the BIPM database. In 2016–2019, 35 CRMs for the composition of dairy and meat products, egg powder, baby food, starch products, oil crops, as well as products on their basis, were created together with the development of primary reference procedures for measuring the mass fraction of fat, crude fat, ash, and carbohydrates. In addition, CRMs for the composition of food additives (glycine, melamine, and cystine) and dairy products were developed to provide metrological support for IR analyzers, as well as a CRM for the composition of reconstituted milk to control measurement results via the enzyme-linked immunosorbent assay (ELISA). The developed CRMs are organized by the application fields of the Technical Regulations of the Customs Union and the sectors comprising the international food triangle

A. S. Sergeeva (✉) · M. Yu. Medvedevskikh
UNIIM – Affiliated Branch of the D. I. Mendeleyev Institute for Metrology, 4 Krasnoarmeyskaya str, Yekaterinburg 620075, Russia
e-mail: sergeevaas@uniim.ru

N. L. Vostrikova
V. M. Gorbatov Federal Research Center for Food Systems of Russian Academy of Sciences, 26 st. Talalikhina, Moscow 109316, Russia
e-mail: n.vostrikova@fncps.ru

S. V. Medvedevskikh et al. (eds.), *Reference Materials in Measurement and Technology*,
https://doi.org/10.1007/978-3-031-06285-8_12

model. These CRMs can be used to ensure uniform measurement of the identification and nutritional value indicators of food products and food raw materials.

Keywords Food products · Reference materials · Measurement standards · Primary reference measurement procedures · Nutritional value indicators · Identification indicator

Introduction

Data on the nutritional value of food products and raw materials are required for performing various tasks: production planning, in-process control, development of new food products, labeling, establishment of product compliance with legal requirements, development of dietary guidelines, menu planning, etc. The indicators of nutritional value include the content of proteins, fats, and carbohydrates. Several food products and raw materials are also subject to requirements for ash content. Water (moisture) content is used as an additional indicator of food quality [1]. When measuring the above indicators, it is necessary to use measurement procedures of specified accuracy while ensuring metrological traceability to a specific comparison basis [2]. Metrological traceability is achieved by using reference materials whose certified values are established using state primary standards and primary reference measurement procedures (PRMPs).

The nomenclature of food products is extensive, thus making it impossible to create a certified reference material (CRM) for each product. This factor prompted the AOAC INTERNATIONAL[1] to develop a food triangle model in the early 1990s [3, 4]. This model allows all food products to be categorized into nine sectors according to their protein, fat, and carbohydrate content; the vertices of the triangle correspond to 100% of each component. For most products allocated to one sector, it is sufficient to study one or two food matrices from that sector to validate an analytical method. This model was adopted by the National Institute of Standards and Technology (NIST,[2] USA) when developing CRMs for the composition of food products and raw materials [5–8]. In the Russian Federation, due to a lack of a unified concept for CRM development, the nomenclature of available CRMs was very limited both in terms of matrices and certified characteristics. In addition, most CRMs were not traceable to state primary standards or PRMPs.

In order to improve metrological support provided for the food industry, over 40 CRMs for the composition of food products, raw materials, and additives were developed by the specialists of the Laboratory for Metrological Support of Moisture Measurement and Reference Materials (UNIIM) from 2008 to 2020. The certified characteristics of these CRMs are traceable to state primary standards and PRMPs. The present article considers the results of CRM development at UNIIM from two

---

[1] AOAC INTERNATIONAL – Association of Analytical Communities, available at: https://www.aoac.org.

[2] NIST – The National Institute of Standards and Technology, available at: https://www.nist.gov/.

Table 1 Nomenclature of CRM materials

Technical regulation	CRM material
TR CU 015/2011 On safety of grain [11]	Wheat, rye, and combined feed
TR CU 021/2011 On food safety [12]	Egg powder, starch, and chocolate
TR CU 024/2011 On safety of fat-and-oil products [13]	Peanut butter, grist, and oil cake
TR CU 033/2013 On safety of milk and dairy products [14]	Milk powder (full-cream, skim), milk-based formula; freeze-dried cottage cheese and sour cream; dried cheese
TR CU 034/2013 On safety of meat and meat products [15]	Freeze-dried beef, pork, and poultry
TR CU 027/2012 On safety of certain types of specialized food products including the therapeutic and preventive dietary food [16]	Milk/milk-free instant rice, buckwheat, and cornmeal porridges for babies
TR EAEU 040/2016 On safety of fish and fish products [17]	Freeze-dried pollock fillet

angles: (1) taking into account the legislative requirements of the Russian Federation in the field of measurement uniformity assurance and technical regulation; (2) in the context of the international food triangle model.

Materials and Methods

The food product nomenclature is so extensive that it is impossible to develop a CRM for each food product. Thus, the first step was to create a nomenclature of CRM materials [9] in continuation of that presented in [10]. Firstly, several stable and homogeneous matrices representing the entire range of products covered by the Technical Regulations of the Customs Union[3] (TR CU) were selected for each effective TR CU (Table 1).

Secondly, the selected CRM materials were allocated to the sectors comprising the food triangle model [3, 4]. Finally, a list of certified characteristics (indicators of nutritional value and identification) was selected, taking into account the available standard and measurement base.

The development of each new CRM included the following stages:

- material preparation (mixing, conditioning, freeze-drying, fraction selection);
- study of long- and short-term stability, homogeneity of the CRM material taking [18] into account;

[3] Effective Technical Regulations of the Customs Union, Rosstandart, available at: https://www.rst.gov.ru/portal/gost/home/standarts/technicalregulationses.

- establishment of the certified value using state primary and secondary standards, as well as PRMPs; estimation of characterization-associated standard uncertainty of the certified value taking [19] into account;
- interlaboratory experiment conducted to verify the applicability of CRMs.

The following primary and secondary standards were used to establish certified values of reference materials at UNIIM:

- GET 173–2017 State Primary Measurement Standard for the units of mass fraction and mass (molar) concentration of water in solid and liquid substances and materials [20];
- GVET 176–1-2010 State Secondary Measurement Standard for the units of mass fraction and mass (molar) concentration of components in solid and liquid substances and materials on the basis of the volumetric titrimetric method [21], which is in turn traceable to GET 176–2019 State Primary Measurement Standard for the units of mass (molar, atomic) fraction and mass (molar) concentration of components in liquid and solid substances and materials on the basis of coulometry.[4]

In addition, the following PRMPs were developed for establishing the certified values of reference materials [22]:

- PRMP for the mass fraction of fat in food products and food raw materials M.241.01/RA.RU.311866/2018[5];
- PRMP for the mass fraction of ash in food products and food raw materials M.241.02/RA.RU.311866/2018[6] [23];
- PRMP for the mass fraction of carbohydrates in food products and food raw materials[7];
- PRMP for the mass fraction of crude fat (oil content) in oilseeds and products on their basis[8] [24].

---

[4] GET 176–2019 State Primary Measurement Standard for the units of mass (molar, atomic) fraction and mass (molar) concentration of components in liquid and solid substances and materials on the basis of coulometry. In: Federal Information Fund for Ensuring the Uniformity of Measurements. https://fgis.gost.ru/fundmetrology/registry/12/items/1382712.

[5] State Primary Reference Measurement Procedure for the mass fraction of fat in food products and food raw materials M.241.01/RA.RU.311866/2018 (FR.PR1.31.2019.00001). In: Federal Information Fund for Ensuring the Uniformity of Measurements. https://fgis.gost.ru/fundmetrology/registry/6/items/595556.

[6] State Primary Reference Measurement Procedure for the mass fraction of ash in food products and food raw materials M.241.02/RA.RU.311866/2018 (FR.PR1.31.2019.00002). In: Federal Information Fund for Ensuring the Uniformity of Measurements. https://fgis.gost.ru/fundmetrology/registry/6/items/595557.

[7] State Primary Reference Measurement Procedure for the mass fraction of carbohydrates in food products and food raw materials (FR.PR1.31.2019.00005). In: Federal Information Fund for Ensuring the Uniformity of Measurements. https://fgis.gost.ru/fundmetrology/registry/6/items/1057023.

[8] State Primary Reference Measurement Procedure for the mass fraction of crude fat (oil content) in oilseeds and products on their basis (FR.PR1.31.2019.00009). In: Federal Information Fund for

Results and Discussion

The first reference materials developed for the food industry were CRMs for the composition of grain, its products, and powdered milk products (Table 2). These CRMs were created as part of activities involving the transfer of water (moisture) content units from GET 173–2008 State Primary Standard for the units of mass fraction and mass concentration of moisture in solid substances and materials [25]. GET 173 developed by the laboratory in 2006–2008 was further improved in 2013 and 2017 [20]. During CRM development, the laboratory applied its accumulated experience in creating and using moisture measurement procedures, estimating measurement uncertainty, and preparing homogeneous and stable material for interlaboratory comparative studies and international comparisons [26, 27].

Further activities of the laboratory were aimed at improving GET 173, developing and certifying measurement procedures, assessing the CRMs for the mass fraction of moisture in terms of their applicability to food quality control [28], as well as providing metrological support for the measurement of nitrogen (protein) mass fraction. In 2010, GVET 176–1-2010 State Secondary Measurement Standards for the units of mass fraction and mass (molar) concentration of components in solid and liquid substances and materials on the basis of the volumetric titrimetric method was approved [21]. Furthermore, as part of GVET 176–1-2010 development, a CRM for the composition of grain and its products (GSO 9734–2010) and a CRM for the composition of milk powder (GSO 9563–2010) were created (Table 3).

GSO 9563–2010 was used in international comparisons within the COOMET project 508/RU/10 (Fig. 1). Following comparisons, GSO 9563–2010 was recognized as an interstate CRM (MSO 1781:2012) to be applied in the territory of the member-states of the Euro-Asian Council for Standardization, Metrology, and Certification.

The mass fraction of nitrogen (protein) in GSO 9563–2010 and GSO 9734–2010 was determined using the Kjeldahl method recognized as an umpire method by

Table 2 CRM having a certified characteristic of moisture mass fraction

CRM name	Range, %	Error, ± Δ, %
GSO 8989–2008 CRM of 1st-category for the mass fraction of moisture in grain	7.0–18.0	0.1
GSO 8990–2008 CRM of 2nd-category for the mass fraction of moisture in grain	7.0–25.0	0.2–0.3
GSO 9564–2010 CRM for the mass fraction of moisture in grain products	7.0–16.0	0.2
GSO 10148–2012 CRM for the mass fraction of moisture in powdered milk products	2.0–10.0	0.08–0.12

Ensuring the Uniformity of Measurements. https://fgis.gost.ru/fundmetrology/registry/6/items/1057069.

Table 3 CRMs having the mass fraction of moisture, nitrogen, and protein as their certified characteristics

CRM name	Certified characteristics	Range, %	Error, ± Δ, %
GSO 9563–2010 CRM for the composition of milk powder (ASM-1) [21]	moisture mass fraction	2.00 – 5.00	0.15
	nitrogen mass fraction*	1.00 – 7.00	0.03
	protein mass fraction*	6.0 – 45.0	0.2
GSO 9734–2010 CRM for the composition of grain and its products	moisture mass fraction	7.0 – 25.0	0.2 – 0.3
	nitrogen mass fraction*	1.00 – 8.00	0.04 – 0.06
	protein mass fraction*	5.00 – 50.00	0.25 – 0.35

* Values are given in terms of dry matter

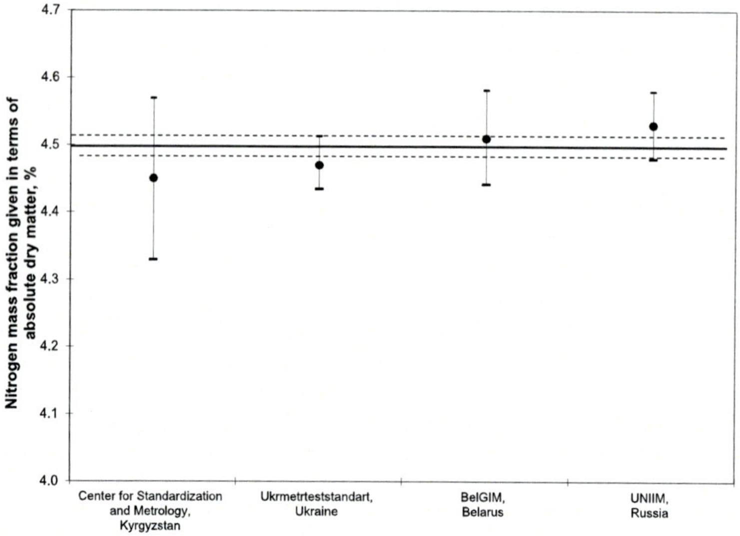

Fig. 1 Results of the COOMET 508/RU/10 comparisons

several international organizations [29]. However, since this method is very labor-intensive and time-consuming, testing laboratories increasingly use rapid measurement methods: Dumas combustion method and IR spectroscopy [30, 31]. These methods require the construction of calibration characteristics using pure substances. In order to provide metrological support for express analyzers, UNIIM developed CRMs for the composition of glycine [32], melamine [33], and cystine (Table 4).

In the following years, work was under way on recognizing the measurement capabilities of UNIIM in the field of nitrogen content measurement at the international level. The following comparisons were carried out: CCQM-K130&P166 key comparisons for determining the mass fraction of nitrogen in pure substance (glycine) [34], as well as pilot and key comparisons for determining the mass fraction of nitrogen in milk powder (CCQM-P167 and CCQM-K149) [35]. The results

Table 4 CRMs for the composition of glycine, melamine, and cystine

CRM name	Certified characteristics	Range, %	Error, ± Δ, %
GSO 10272–2013 CRM for the composition of glycine [32]	nitrogen mass fraction	18.47–18.66	1.0 rel%
	mass fraction of the main substance	99.0–100.0	1.0 rel%
GSO 10825–2016 CRM for the composition of melamine [33]	nitrogen mass fraction	63.30–66.64	1.5 rel%
	mass fraction of the main substance	95.0–100.0	1.5 rel%
GSO 11337–2019 CRM for the composition of cystine	carbon mass fraction	29.8–30.1	0.3 abs%
	hydrogen mass fraction	4.9–5.1	0.2 abs%
	nitrogen mass fraction	11.5–11.8	0.2 abs%
	sulfur mass fraction	26.5–26.8	0.3 abs%

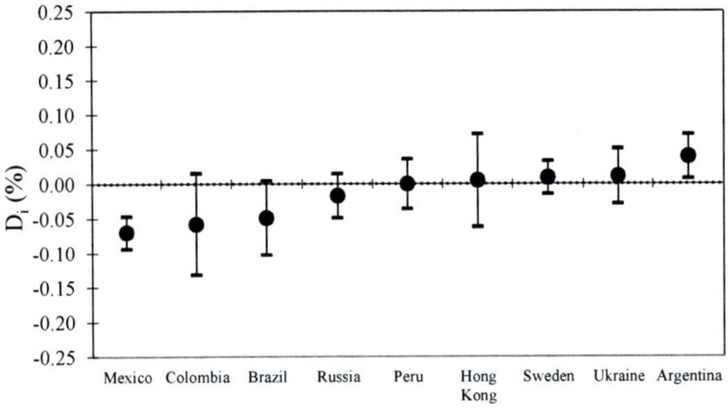

Fig. 2 Results of the CCQM-K149 comparisons determining the mass fraction of nitrogen in milk powder [35]

of the CCQM-K149 key comparisons using GSO 9563–2010 are shown in Fig. 2. The successful participation of UNIIM in comparisons led to the publication of 6 CMC[9] lines for the measurement of nitrogen mass fraction in glycine, milk powder, grain, egg powder, porridge, and feed in the BIPM[10] database.

In 2016–2019, the laboratory carried out work on the development, approval, and certification of PRMPs to establish the operationally determined indicators of nutritional value: mass fractions of fat, crude fat (oil content), carbohydrates, and ash. As a result, thirty-five CRMs for the composition of food products and food raw materials were developed using PRMPs: dairy and meat products, egg powder, baby food, oilseeds and products on their basis, as well as starch products [36]. Here, it should be noted that the CRM development was carried out in close cooperation

[9] CMC – calibration and measurement capabilities.

[10] BIPM - International Bureau of Weights and Measures. https://www.bipm.org.

with other organizations, specifically food industry specialists. For instance, V. M. Gorbatov Federal Research Center for Food Systems (RAS, Moscow) was the co-developer of GSO 11274–2019/GSO 11276–2019 for the composition of freeze-dried meat products [37]. As an example, Table 5 provides the metrological characteristics of GSO 11086–2018/GSO 11091–2018 CRMs for the composition of powdered milk products (ASM-2 CRM set UNIIM) and GSO 11399–2019 CRM for the composition of milk powder (ASM-3 CRM UNIIM).

The applicability of GSO 11086–2018/GSO 11091–2018 was confirmed by the results of interlaboratory comparisons (ILCs) 241-MP3 for determining the quality indicators of milk and dairy products. Table 6 summarizes the results of the ILC round 241-MP3-3, demonstrating the applicability of these CRMs to control the accuracy of measured moisture, protein, and fat mass fraction values.

Table 5 Metrological characteristics of GSO 11086–2018/GSO 11091–2018 and GSO 11399–2019

CRM name	Certified characteristics	Range, %	Error, ± Δ, %
GSO 11086–2018/GSO 11091–2018 CRMs for the composition of powdered milk products (ASM-2 CRM set UNIIM)	moisture mass fraction	2.00–10.00	0.08–0.12
	nitrogen mass fraction*	0.20–11.00	0.03
	protein mass fraction*	1.2–70.0	0.2
	fat mass fraction*	0.10–80.00	0.10–0.25
GSO 11399–2019 CRM for the composition of milk powder (ASM-3 CRM UNIIM)	lactose mass fraction*	30.0–55.0	5.0
	mass fraction of carbohydrates*	30.0–70.0	0.4

* Values are expressed in terms of dry matter

Table 6 Summarized results of the ILC round 241-MP3-3/2018 *determination of quality indicators of milk and dairy products*

CRM code	Parameter to be determined	Percentage of laboratories reporting satisfactory results	Percentage of laboratories reporting indeterminate results	Percentage of laboratories reporting unsatisfactory results
MP3(1)	moisture mass fraction	100	0	0
	protein mass fraction	84	11	5
	fat mass fraction	78	18	4
MP3(2)	moisture mass fraction	93	7	0
	protein mass fraction	90	0	10
	fat mass fraction	77	0	23

In 2020, CRMs for the composition of milk and light cream were developed to provide metrological support for rapid IR analyzers. The certified mass fraction values of dry matter, protein, fat, and lactose were established using GET 173–2017 and GVET 176–1-2010, as well as certified measurement procedures developed to expand the scope PRMPs (FR.PR1.31.2019.00001 and FR.PR1.31.2019.00005). Additional measurements were performed at the testing laboratory of the Ural State University of Economics (Yekaterinburg) employing standardized measurement procedures [38].

A new area of focus for the laboratory consists in providing metrological support for the enzyme-linked immunosorbent assay (ELISA). Thus, GSO 11168–2018 CRM for the composition of reconstituted milk (RM CRM UNIIM) was developed in collaboration with Chema LLC (Moscow) in 2018 [39]. The certified values of this CRM include the mass fraction of nitrogen and the mass concentration of milk powder.

To date, the Laboratory for Metrological Support of Moisture Measurement and Reference Materials produces over 40 CRMs for the composition of food products, food raw materials, and food additives; these CRMs can be used to provide metrological support for measuring the identification indicators of food products in order to ascertain their compliance with the TR CU requirements (Fig. 3).

The produced CRMs cover seven of the nine sectors comprising the international food triangle model [3, 4] (Fig. 4). Thus, they can be used to validate analytical methods for food products and food raw materials from these sectors.

In the near future, the laboratory plans to develop CRMs for the remaining sectors of the food triangle and product groups specified in the Technical Regulations: CRMs for the composition of freeze-dried fish, chocolate, peanuts, soy flour, and sugar. The

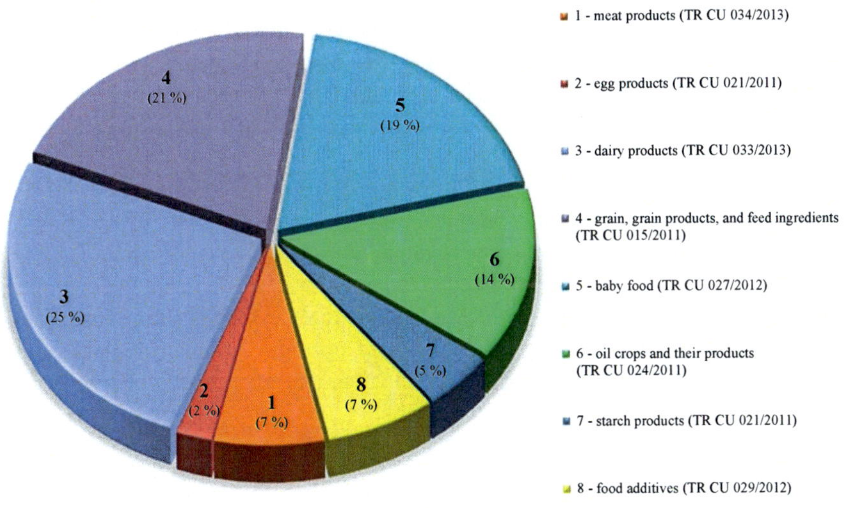

Fig. 3 Distribution of produced CRMs for the composition of food products depending on items governed by TR CU

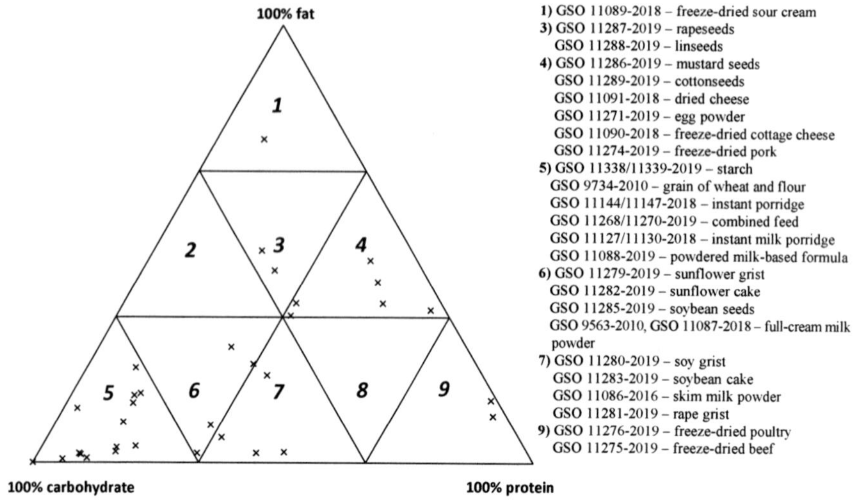

Fig. 4 Distribution of the produced CRMs among the sectors comprising the food triangle

next step would be to expand the certified characteristics of the developed CRMs by adding information on carbohydrate, fatty acid, and amino acid compositions.

Conclusion

In 2008–2020, the specialists of the Laboratory for Metrological Support of Moisture Measurement and Reference Materials (UNIIM) developed a complex designed to provide metrological support for measuring the nutritional value of food products, including state standards for measurement units (GET 173–2017 and GVET 176–1-2010), PRMPs, and over 40 CRMs for the composition of food products and food raw materials.

The developed CRMs ensure uniformity in measuring the identification and nutritional value indicators of food products and food raw materials to ascertain the compliance of products with technical regulation laws, as well as the consistency between measurement results obtained in the territory of the Russian Federation and internationally. Furthermore, the CRM nomenclature developed taking into account the international food triangle model helps to optimize the metrological support system of the food industry by using a limited number of matrices for a wide range of analyzed objects.

Author Contributions Anna S. Sergeeva: definition of the idea and methodology of the article; collection and analysis of literary data (including in foreign languages); preparation of the first draft of the article; writing, formatting, and revision of the text.

Natalia L. Vostrikova: supervision of the experimental studies; critical analysis of the article.

Maria Yu. Medvedevskikh: concept development; study initiation; definition of the idea and methodology of the article; supervision of the experimental studies; critical analysis of the article.

Conflict of Interest The article was prepared on the basis of a report presented at the IV International Scientific Conference "Reference Materials in Measurement and Technology" (St. Petersburg, December 1–3, 2020). The article was admitted for publication after the abstract was revised, the article was formalized and the review procedure was carried out.

The version in the Russian language is published in the journal "Measurement Standards. Reference Materials" 2021;17(1):23–33. https://doi.org/10.20915/2687-0886-2021-17-1-23-33.

References

1. Nielsen SS (2017) Food analysis, 5th edn. Springer, New York, p 649
2. ISO/IEC 17025 (2017) General requirements for the competence of testing and calibration laboratories. ISO, Geneve
3. Ikins W, DeVries J, Wolf WR, Oles P, Carpenter D, Fraley N et al (1993) A food matrix organizational system applied to collaborative studies. The Referee AOAC Inter 17(7):1–7
4. Wolf WR, Andrews KW (1995) A system for defining reference materials applicable to all food matrices. The Referee AOAC Inter 352(1–2):73–76. https://doi.org/10.1007/BF00322300
5. Sharpless KE, Welch MJ, Greenberg RR, Iyengar GV, Colbert JC (1998) Recent SRMs for organic and inorganic nutrients in food matrices. Fresenius J Anal Chem 360(3–4):456–458. https://doi.org/10.1007/s002160050738
6. Sharpless KE, Thomas JB, Christopher SJ, Greenberg RR, Sander LC, Schantz MM et al (2007) Standard reference materials for foods and dietary supplements. Anal Bioanal Chem 389(1):171–178. https://doi.org/10.1007/s00216-007-1315-y
7. Phillips MM, Sharpless KE, Wise SA (2013) Standard reference materials for food analysis. Anal Bioanal Chem 405(13):4325–4335. https://doi.org/10.1007/s00216-013-6890-5
8. Wise SA, Phillips MM (2019) Evolution of reference materials for the determination of organic nutrients in food and dietary supplements-a critical review. Anal Bioanal Chem 411(1):97–127. https://doi.org/10.1007/s00216-018-1473-0
9. Medvedevskikh MY, Krasheninina MP, Sergeeva AS, Nikonova NA (2018) On the development of reference materials of food quality indicators. In: *Reference materials in measurement and technology: Conference abstracts III International scientific conference, 11–14 September* 2018. Yekaterinburg, Russia. UNIIM, Yekaterinburg, pp 129–131 (in Russ)
10. Zaporozhets AS, Parfenova EG, Gushchina MO (2013) Metrological support of product moisture measurement (within the framework of the requirements of Customs Union Technical Regulations). Conformity Assess Methods 9:19–22 (In Russ)
11. TR CU 015/2011 On safety of grain: Technical Regulations of the Customs Union. https://ec.europa.eu/food/sites/food/files/safety/docs/ia_eu-ru_sps-req_decision-874_en.pdf
12. TR CU 021/2011 On food safety: Technical Regulations of the Customs Union. https://en.cyclopedia.ifcg.ru/wiki/Technical_Regulation_of_the_Customs_Union_021/2011_%22On_food_safety%22
13. TR CU 024/2011 On safety of fat-and-oil products: Technical Regulations of the Customs Union. https://en.cyclopedia.ifcg.ru/wiki/Technical_Regulation_of_the_Customs_Union_024/2011_%22On_safety_of_fat-and-oil_products%22
14. TR CU 033/2013 On safety of milk and dairy products: Technical Regulations of the Customs Union. https://en.cyclopedia.ifcg.ru/wiki/Technical_Regulation_of_the_Customs_Union_033/2013_%22On_safety_of_milk_and_dairy_products%22

15. TR CU 034/2013 On safety of meat and meat products: Technical Regulations of the Customs Union. https://en.cyclopedia.ifcg.ru/wiki/Technical_Regulation_of_the_Customs_Union_034/2013_%22On_safety_of_meat_and_meat_products%22
16. TR CU 027/2012 On safety of certain types of specialized food products including the therapeutic and preventive dietary food: Technical Regulations of the Customs Union. https://ec.europa.eu/food/sites/food/files/safety/docs/ia_eu-ru_sps-req_decision-34_annex_en.pdf
17. TR EAEU 040/2016 On safety of fish and fish products: Technical Regulation of the Eurasian Economic Union. https://en.cyclopedia.ifcg.ru/wiki/Technical_Regulation_of_the_Eurasian_Economic_Union_040/2016_%22On_safety_of_fish_and_fish_products%22
18. ISO Guide 35 (2017) Reference materials – Guidance for characterization and assessment of homogeneity and stability. ISO, Geneve
19. Ellison SLR, Williams A (eds) (2012) Eurachem/CITAC guide: quantifying Uncertainty in Analytical Measurement. 3th ed. https://www.eurachem.org/images/stories/Guides/pdf/QUAM2012_P1.pdf
20. Medvedevskikh MY, Sergeeva AS, Krasheninina MP, Shokhina OS (2019) Creating reference standards for the implementation of the State verification scheme for water content measurement. Meas Tech 62(6):475–483. https://doi.org/10.1007/s11018-019-01649-3
21. Krasheninina MP, MYu, Medvedevskikh, Medvedevskikh SV, Sobina EP, Neudachina LK (2013) An estimate of the metrological characteristics of a standard sample of the composition of dried whole milk using primary and secondary state standards. Meas Tech 56(9):1076–1082. https://doi.org/10.1007/s11018-013-0333-8
22. Sergeeva AS, Medvedevskikh MY, Studenok VV, Parfenova EG (2020) Use of primary reference measurement techniques in the field of assurance of measurement uniformity. Zakonodatel'naya i prikladnaya metrologiya = Legal and Applied Metrology 5(167):5–9 (In Russ)
23. Medvedevskikh MYu, Sergeeva AS, Krasheninina MP, Shokhina OS (2019) State primary reference procedure for the measurement of ash mass fraction in food, foodstuff and alimentary raw materials. Ind Labor Diagn Mater 85(6):70–80. https://doi.org/10.26896/1028-6861-2019-85-6-70-80 (In Russ)
24. Sergeeva AS, Parfenova EG, Golynets OS (2020) Development of primary reference measurement procedure and reference materials for mass fraction of crude fat (oil content) in oilseeds and other products on their base. Measur Stand Ref Mater 16(3):37–51. https://doi.org/10.20915/2687-0886-2020-16-3-37-51 (In Russ)
25. Gorshkov VV, Koryakov VI, MYu, Medvedevskikh, Medvedevskikh SV (2010) State primary standard of unit of mass fraction and unit of mass concentration of moisture in solid substances and solid fabricated materials. Meas Tech 53(4):386–390. https://doi.org/10.1007/s11018-010-9515-9
26. Medvedevskikh SV, Medvedevskikh MY, Karpov YA (2015) General approaches to the estimation of uncertainty in the results of reproducing units of water content in solids and materials. Meas Tech 58(8):926–933. https://doi.org/10.1007/s11018-015-0819-7
27. Medvedevskikh SV, Medvedevskikh MY, Krasheninina MP (2020) Quality assessment of the preparation of reference materials of moist solid substances. In: Medvedevskikh S, Kremleva O, Vasil'eva I, Sobina E (eds) Reference Materials in Measurement and Technology. RMMT 2018. Switzerland, Springer, Cham, pp 153–173. https://doi.org/10.1007/978-3-030-32534-3_16
28. MYu, Medvedevskikh, Kazennova NK, Krasheninina MP (2013) Certification of measurement procedures and applicability evaluation of reference materials of moisture mass fraction for quality control of food products in the course of their production, processing and storage. Refer Mater 4:40–45 (In Russ)
29. Moore JC, DeVries JW, Lipp M, Grifiths JC, Abernethy DR (2010) Total protein methods and their potential utility to reduce the risk of food protein adulteration. Compreh Rev Food Sci Food Saf 9(4):330–351. https://doi.org/10.1111/j.1541-4337.2010.00114.x
30. Ingle PD, Christian R, Purohit P, Zarraga V, Handley E, Freel K, Abdo S (2016) Determination of protein content by NIR spectroscopy in protein powder mix products. J AOAC Int 99(2):360–363. https://doi.org/10.5740/jaoacint.15-0115

31. Maehre HK, Dalheim L, Edvinsen SK, Elvevoll EO, Jensen I-J (2018) Protein determination—methods matter. Foods 7(1):5. https://doi.org/10.3390/foods7010005
32. Krasheninina MP, Neudachina LK, Sergeeva AS, Sobina EP (2015) Creation of certified reference material based on glycine. Ref Mater 1:23–33 (In Russ)
33. Krasheninina MP, Makarova SG, Shokhina OS (2017) Development of high-precision method for determination of nitrogen and basic substance in melamine by the Kjeldahl method. Indus Labor Diagn Mater 83(6):64–71 (In Russ)
34. Medvedevskikh M, Krasheninina M, Rego ECPD, Wollinger W, Monteiro TM, Carvalho LJD et al (2017) Final report of CCQM-K130 nitrogen mass fraction measurements in glycine. Metrologia 54(1A):08004. https://doi.org/10.1088/0026-1394/54/1A/08004
35. Medvedevskikh M, Krasheninina M, Cárdenas IAG, Erazo LVM, Forigua DAA et al. (2019) Report of CCQM-K149 nitrogen mass fraction measurements in milk powder. Metrologia 56(1A):08011. https://doi.org/10.1088/0026-1394/56/1A/08011
36. Medvedevskikh MYu, Sergeeva AS (2020) Problems of ensuring metrological traceability of measurement results of indicators of quality for food products and food raw materials. Measur Techn 3:64–70. https://doi.org/10.32446/0132-4713.2020-3-64-70 (In Russ)
37. MYu, Medvedevskikh, Sergeeva AS, Krasheninina MP, Vostrikova NL, Semenova AA, Kuznetsova OA (2019) About the development of reference materials of meat and meat product composition. J Phys: Conf Ser 1420:012030. https://doi.org/10.1088/1742-6596/1420/1/012030
38. Medvedevskikh MYu, Sergeeva AS, Kasilyunas AV, Shatskikh EV, Kolberg NA (2021) Certified reference materials of dairy products composition for IR milk analyzers verification. Food Ind 1:16–19. https://doi.org/10.24411/0235-2486-2021-10003 (In Russ)
39. Medvedevskikh MYu, Medvedevskikh SV, Sergeeva AS, Petukhov PA (2020) Methods of identification and quality control of milk and dairy products: development and application in the field of technical regulation. Zakonodatel'naya i prikladnaya metrologiya = Legal and Applied Metrology 4(166):13–16 (In Russ)

Anna S. Sergeeva Cand. Sci. (Chem.), Senior Researcher of the Laboratory for Metrological Support of Moisture Measurement and Reference Materials, UNIIM – Affiliated Branch of the D. I. Mendeleyev Institute for Metrology.

Natalia L. Vostrikova Dr. Sci. (Eng.), Head of the Research and Development Center, V.M. Gorbatov Federal Research Center for Food Systems of Russian Academy of Sciences.

Maria Yu. Medvedevskikh Cand. Sci. (Eng.), Head of the Laboratory for Metrological Support of Moisture Measurement and Reference Materials, UNIIM – Affiliated Branch of the D. I. Mendeleyev Institute for Metrology.

Features of the "Plants" Cluster in the Collection of Reference Materials of Vinogradov Institute of Geochemisty SB RAS

Elena V. Shabanova, Irina E. Vasil'eva, Dmitrii S. Tausenev, Stefanie Scherbarth, and Uwe Pierau

Abstract The collection of reference materials (RMs) developed at the Vinogradov Institute of Geochemistry of the Siberian Branch of RAS includes four multi-element RMs comprising the "Plants" cluster. These RMs include parts of terrestrial plants (birch leaves, pine needles, mown meadow grass) and the aquatic plant Elodea Canadensis (roots, stems, leaves, and flowers). The plants, which are sensitive indicators of environmental conditions, were collected from unpolluted territories of Eastern Siberia (both near and from Lake Baikal). Differences in the methods of selection and preparation of the material are described. The characteristics of these RMs, such as granulometric composition (shape, size, and particle size distribution), homogeneity, and minimum representative sample weight, as well as the stability of powders under natural aging conditions, were studied in accordance with the Russian and international requirements using modern instruments and methods of chemical analysis. The elemental compositions of plant samples, estimated using the method of interlaboratory certification, are represented by more than 60 elements, from 23 to 41 of which have been certified. The traceability of the results was ensured by the participation of 20 to 38 accredited laboratories from Russia and other countries, as well as the use of more than 10 different analytical methods. A comparison of the certified plant-matrix RMs developed by IGC SB RAS with China plant-matrix CRMs demonstrated their consistency. Following the discussion of their characteristic properties, four plant-matrix RMs are recommended for chemical measurements when validating existing and developing new analytical methods, for quality

E. V. Shabanova (✉) · I. E. Vasil'eva
Vinogradov Institute of Geochemistry, Siberian Branch of the Russian Academy of Sciences (IGC SB RAS), 1A Favorsky str., Irkutsk 664033, Russia
e-mail: shev@igc.irk.ru

I. E. Vasil'eva
e-mail: vasira@igc.irk.ru

D. S. Tausenev
OOO "Sympatec", 51 8th Marta str., Yekaterinburg 620142, Russia

S. Scherbarth · U. Pierau
Sympatec GmbH, 1 Am Pulverhaus, 38678 Clausthal-Zellerfeld, Germany
e-mail: lab@sympatec.com

S. V. Medvedevskikh et al. (eds.), *Reference Materials in Measurement and Technology*,
https://doi.org/10.1007/978-3-031-06285-8_13

control and assessment of the traceability of results when determining a wide range of elements in plant materials, as well as for professional testing of geoecological, pharmaceutical, and agricultural laboratories.

Keywords Multi-element plant reference materials · Granulometric composition of powders · Particle form and sizes · Homogeneity and representative mass of samples · Substance stability and estimated expiration date · Interlaboratory certification · Evaluation of element composition and traceability of measurement results

Introduction

The species diversity of plants is enormous, but the list of existing multi-element plant-matrix reference materials (RM) in Russia is rather small. Vinogradov Institute of Geochemistry of the Siberian Branch of the Russian Academy of Sciences, Russia (IGC SB RAS) has been creating and developing a collection of RMs for more than 40 years. Currently, the collection includes multi-element RMs of the composition of natural and man-made environments, which contain 5–70 certified and recommended elements and components [1]. The collection focuses on ensuring the uniformity of measurements in geoanalysis, representing an important link in interdisciplinary research into fundamental relationships between natural objects, as well as in assessing the state of the environment. The section "Biological Reference Materials" of the collection includes a cluster of "Plants". This cluster is represented by three RMs certified in 2007 and one RM developed in 2018 [2]:

1. RM of the composition of birch leaf (LB-1) GSO 8923–2007/RM COOMET 0067–2008-RU (*Birch leaf*—LB-1),
2. RM of the composition of a mixture of meadow herbs (Tr-1) GSO 8922–2007/RM COOMET 0066–2008-RU (*Mixture of meadow herbs*—Tr-1),
3. RM of the composition of *Elodea canadensis* (EK-1) GSO 8921–2007/RM COOMET 0065–2009-RU (*Canadian pond weed*—EK-1),
4. RM of the composition of Siberian pine needles (HSS-1) (*Pinus Sylvestris*—HSS-1).

The developed RMs are intended to verify analytical methods, to ensure the correctness and traceability of measurement results of chemical elements in plants by various analytical methods when performing geochemical and ecological research projects. The listed plants are sensitive indicators of pollution of natural terrestrial and aquatic ecosystems. The plants for creating the RMs were selected based on such their specific features as: the ability to rapidly change elemental composition under the conditions of environmental pollution; a wide area of growth; the possibility of collecting the amount of biomass sufficient for preparing an RM. The material for the RMs was collected across areas with a minimum human impact. The developed RMs comprising the "Plant" cluster are characterized by the completeness of elemental composition: 63–68 chemical elements. The number of certified, recommended, and informative elements in different samples varies across the range of 32–41, 2–16, and 6–34, respectively.

In this article, we analyze the element composition and some properties of the RMs from the "Plant" cluster of the IGC SB RAS collection to evaluate their traceability to RMs of other producers.

Materials and Methods

An RM is a "material, sufficiently homogeneous and stable concerning one or more specified properties, which has been established to be fit for its intended use in a measurement process" [3]. It is important to know and consider specific properties of an RM substance for which one or more specified properties are characterized by a metrologically sound procedure, specified in the RM certificate, providing the value of the specified property, the associated uncertainty and the statement of metrological traceability.

When producing plant-matrix RMs, it is necessary to consider such features as a different structure and hardness of stems, leaves, and other plant organs, as well as the instability of organic matter because of the possibility of its decomposition during prolonged storage. When the substance is ground and stored, these properties affect the shape, size, and particle-size distribution of RM powders and, consequently, the homogeneity of the RM, the value of the representative analytical mass, and the shelf life of the substance. The development of plant-matrix RMs of the "Plant" cluster relied on regulatory documents for characterizing and estimating the substance homogeneity and stability [4–7]. However, as noted previously [8], the choice of analytical procedures for material selection and preparation, as well as statistical data processing, was determined by the specific properties of matrices and the analyte level in each RM. Therefore, nonstandard analytical methods and approaches were required.

Selecting and Preparing Materials for RMs

The required mass of plant material for each RM of the "Plants" cluster in the IGC SB RAS collection was selected in different years.

The material for GSO LB-1 was collected in a birch grove on the outskirts of the city of Irkutsk, far from actively used roads in the period of mass leaf fall. During dry sunny weather, in the autumn of 1993, the upper layer of fallen birch leaves was collected in clean tarpaulin bags and air dried at room temperature (22 ± 1 °C) in a laboratory.

The initial natural material for GSO Tr-1 was a meadow grass mixture mowed in the summer of 1995 in a natural meadow near the Khomutovo village (Irkutsk Oblast).

The material was processed into a granulated grass powder at an automated grass mill "Travyany" (Russia), packed into paper bags and transported into a laboratory.

Samples of *Elodea Canadensis algae* were collected in the southern part of Mukhor Bay (Maloe Sea, Lake Baikal) during the summer periods of 1999, 2001, and 2002. The material was collected from an inflatable boat. The collected algae were sorted out from other aquatic plants, washed clean of mineral impurities, spread on a tarpaulin, and air-dried under shaded conditions.

The material for the HSS-1 RM was prepared from the needles of *Pinus Sylvestris* (Siberian pine). The needles were cut off from their resinous base from the branches of pine trees of different age in pine groves of the Baikal region, including the Olkhon Island (Irkutsk Region) and the Tunkinskaya Valley (Buryatia), during the summer periods of 2005–2008. The material was air dried in a ventilated laboratory room under shaded conditions.

The method used for preparing an RM affects its preservation (without degradation) over a long period of time; therefore, the air-dried samples were additionally dried in a desiccator at about 70 °C. Radiation sterilization of the original or homogenized material was not performed. The LB-1 and EK-1 materials were cut into fragments $\leq$ 1 mm with a cutter and then was fine-treated in an MEK-50 (Germany) ball mill with a porcelain header to a particle size of less than 0.14 mm. Pellets Tr-1 and pine needles, cut into fragments less than 1 mm on a Pulverisette 15 (Fritsch, Germany) cutting mill, were additionally ground on a MBL-50 (Russia) ball mill with a steel headset. The crushed substance was sieved through a nylon sieve (cell 0.14 mm) and homogenized on a rotating mixer table for 10–15 cycles. The substance was kept for several days and packed in polyethylene jars with a capacity of 50, 100, and 150 cm^3.

Granulometric Composition of Powders: Shape, Size, and Particle Size Distribution

The initial assessment of the granulometric composition of powders of three samples LB-1, Tr-1, and EK-1 was conducted by dry screening on a standard set of five Fritsch sieves (Germany) with an additional air suction using a low-power household vacuum cleaner. The granulometric composition of the HSS-1 material was assessed by dry screening using a set of Retsch sieves (Germany) with a mesh size of 0.16–0.04 mm, when the powder was pushed with a plastic brush. The complexities of sieving were due to the fibrous structure of the plant materials, the electrostatic sticking of particles when using the air-dry sieving method and the inability to use sieving in an aqueous environment due to the swelling of the particles. Therefore, the distribution of powders in terms of coarseness, form, and particle size for 4 RMs was additionally investigated on a dynamic analyzer of images Sympatec QICPIC (an optical range M7 of 4.2–2,888/8,665 μm) with an air dispersion module RODOS/L (Sympatec GmbH, Germany). In order to decrease the influence of the statistical error, the samples were divided into equal portions with the aid of a revolving divider. The Sympatec QICPIC & RODOS/L system is based on the physical principle of a dynamic image analysis, which measures the particle size and shape distribution, as well as characterizes the shape of individual particles by the given shape parameters. Distributions in terms of particle size were performed using calculation variants of the area of equivalent circle and Feret diameter (minimum/maximum). Shape factor distribution diagrams for aspect ratio and sphericity (by particle size and in aggregate) were plotted, suggesting the shape distribution of the sample material.

For each RM, measurements were performed for three (portions) jars. During dispersion, 4.80 g (3 × 1.60 g) of powder mass from each jar was examined at 0.2 bar. In addition to particle size distribution, the description of particle shapes was given in several ways. The shape factor distribution diagrams for aspect ratio and sphericity (by particle size and in aggregate) suggest the distribution and shape of individual powder particles.

Homogeneity of Plant Materials

The homogeneity of each sample was assessed by a single-factor analysis of variance for 20 randomly selected jars (20 × 3) according to [4, 7] in the laboratories of the Center for Collective Use "Isotope-Geochemical Research" at the IGC SB RAS. The experimental data for evaluating the homogeneity of the LB-1, Tr-1, and Ek-1 samples were obtained using X-ray fluorescence spectrometry. The measurements were taken using an X-ray spectrometer S4 Pioneer (Bruker, Germany) for a 1 g sample. Estimating the error from heterogeneity required relative spectral line strengths of 8, 15 and 16 elements, respectively. The homogeneity of the HSS-1 powder was studied using the method of X-ray fluorescence spectrometry for 15 elements from the 0.5 g sample. The spectral line intensities of 25 elements were also measured on an atomic emission spectrometer with inductively coupled plasma iCAP 6300 Duo (Thermo Scientific, USA) after autoclave acid decomposition of 0.15 and 0.1 g samples. When estimating the error from the heterogeneity of the elements, the intensity changes of which could not be calculated due to the small element contents, this characteristic was attributed the maximum value set for the indicator element.

Characteristics of Plant Materials

Certification of a large number of analytes in plant RMs required a strategy for an interlaboratory comparison [4–7]. The chemical elements that can be certified in a substance were determined by both the primary methods of analysis, including gravimetry (GM), titrimetry (TM) and instrumental neutron activation analysis (INAA), and atomic spectroscopy. The latter methods require calibration by an RM of an adequate composition, such as: inductively coupled plasma atomic emission spectroscopy (ICP-AES), atomic emission spectroscopy with arc discharge (AES with arc), laser-induced breakdown spectroscopy (LIBS), flame atomic emission spectroscopy (FAES); flame and electrothermal atomic absorption spectrometry (FAAS and ET-AAS); inductively coupled plasma mass spectrometry (ICP-MS); X-ray fluorescence spectrometry (XRF); spectrophotometry (SPM); near-infrared spectrometry (NIRS). The certified mass fractions of elements in plant-matrix RMs were established based on the results of 10–12 analytical methods with different chemical and physical bases using different methods of transferring samples to solutions.

The traceability of the measurement results obtained within the framework of an interlaboratory comparison to an SI unit [7] was realized through the participation of more than 20 competent testing laboratories in Russia and other countries (including those accredited for compliance with GOST ISO/IEC 17,025 [9]), which were used in the analysis of verified measuring instruments. To control the accuracy of the measurement results, we used RMs with established metrological traceability.

Stability of Plant Matrix RMs

To substantiate the shelf life of RMs, we applied a method that involves establishing the dependence of the change in the certified value depending on the time under natural aging conditions [10]. Initial studies of the stability of plant-matrix RM LB-1, Tr-1, and EK-1, performed in 2002–2007 for a limited number of chemical elements, guaranteed a 5-year shelf life. Similar studies of 2008–2013 confirmed the stability of certified RMs for 10 years when stored indoors at room temperature.

The shelf life of plant RMs was adjusted according to the quality control of measurements using different methods of chemical analysis in 22 accredited testing laboratories in Russia, Spain, Belarus, Kazakhstan, and Mongolia in 2008–2017. The interlaboratory comparison to study the stability of a substance was planned according to [4, 10] and carried out using the ICP-AES, FAES, SAES, LIBS; FAAS and ET-AAS; ICP-MS; XRF; GM; and SPM methods. The largest amount of experimental data was obtained using multi-element analysis techniques (ICP-AES, ICP-MS, XRF). Prior to combining the results obtained by different methods into a general sample, their homogeneity was assessed, and the inconsistent data were excluded. Based on the minimum number of measurements required to assess the stability of an RM, one year was taken as the time interval between the measurements. If the content of a particular element was determined several times during a year, the results obtained during each year were averaged [10]. Variations in the content of 35, 38, and 40 chemical elements in the EK-1, Tr-1, and LB-1 RM, respectively. Calculations of the measured contents of the certified elements showed that the dependence between the measurement uncertainty and the storage time was not statistically significant. The plant RMs prepared according to the above procedure can be used for their intended purpose for 20 years and longer, given they are stored in a laboratory at room temperature. For chemical elements, the contents of which were set roughly, experimental data were not provided. Therefore, we used only literature data on the natural stability of the corresponding biological compounds. It concerns the stability of the HSS-1 material; since it was prepared in a similar way for the three matrix RMs. Monitoring over the stability of RMs included in the "Plants" cluster is currently underway.

Results and Discussion

Particle Shape and Particle Size Distribution, Granulometric Composition

The quantitative elemental composition and homogeneity of the distribution of each element in the original and prepared RM substance are individual. These characteristics reflect such specific features of a plant material (RM candidate) as species affiliation; the structure of the plant's organs, which are selected for the RM; the composition of soil and the climatic conditions for growth; the cycle of the plant's development; the time of material collection for the RM; preparation methods, etc. Each element is often represented by several organoelemental compounds, which are unevenly distributed between the organs and tissues of the plant. In cases where homogeneity in a certain mass of a powder is not be achieved during crushing and homogenization, this will increase the error of a single analytical procedure, eventually increasing the uncertainty (error) of the certified value.

Examination of the plant powders under study by an energy-dispersion X-ray microanalyzer Phenom-ProX (B. V. Phenom-World, the Netherlands) showed the presence and wide variety of conditionally spherical and nonspherical particles of different sizes (Fig. 1). A dynamic image analysis allowed the number of particles of the real shape to be controlled. A comparison of the data on particle size distribution obtained by sieving and by a Sympatec QICPIC & RODOS/L dynamic image analyzer indicated a significant difference between the distributions obtained when forcing plant powders through sieve cells or when sucking them into the measuring channel with an air jet (Table 1).

The Sympatec QICPIC & RODOS/L analytical system determined from 8,069,830 to 24,120,270 particles in different plant samples per 1 g of substance. All the samples contain both spherical and elongated particles of various shapes (Fig. 2). An evaluation of the shape, size and distribution of particles in the LB-1, Tr-1, and HSS-1 RMs showed a decrease in the sphericity (the ratio of the perimeter of an equivalent circle to the actual perimeter) of organic matter particles when increasing their size. Large particles of EK-1 are characterized by an increasing value of this indicator. All RMs are characterized by a tendency for the aspect ratio to decrease with increasing particle size, i.e. "stretching" was observed for larger particles.

A comparison of the grinding degree of four RMs indicated a wide range of particle sizes, typical for all powders (Fig. 3). The Tr-1 powder is represented by the

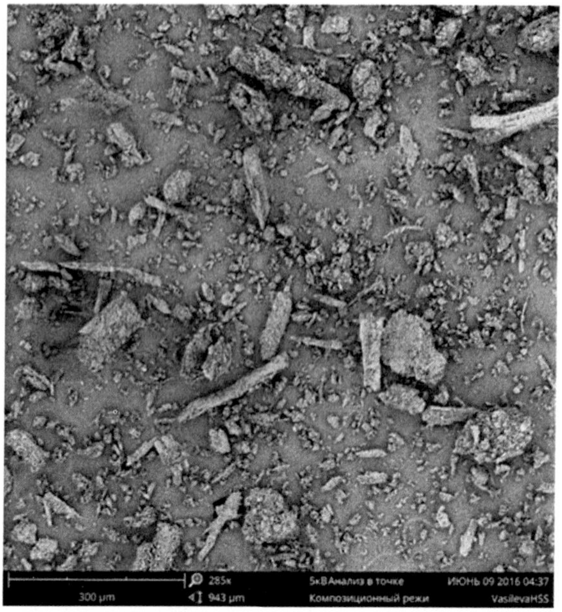

Fig. 1 A powder from Siberian pine needles

smallest particles (average size 20–30 μm), and EK-1 powder is represented by the largest particles (80–110 μm). For HSS-1 and LB-1, this indicator was 50–80 and 20–30 μm, respectively. The largest variations of particle size were observed for the LB-1 powder, which is characterized by the largest proportion of particles with a size of 100–500 μm.

The obtained results indicate that the used method of grinding plants produces completely homogeneous materials and that small masses of such powders, selected for the analysis, can show good reproducibility of repeated measurements when no systematic errors prevent the quantitative measurement of the contents. At the same time, the repeated measurements of the granulometric composition of plant materials showed the presence of a changing number of large particles in the powder; therefore, an analytical portion should be larger than the representative mass.

It is important to note that the use of a dynamic image analyzer has provided a new opportunity to select RM powders to the studied samples of plants similar not only in terms of chemical composition, but also in terms of structure and size. Undoubtedly, the data on particle size distribution obtained using the Sympatec QICPIC & RODOS/L dynamic image analyzer are more realistic than those obtained by forced sieving, since the former considers different projections of particles in motion.

Table 1 Results of granulometric analysis of plant RMs

Fraction, μm	Mass fraction, %							
	LB-1		Tr-1		EK-1		HSS-1	
	Sieving	QICPIC/RODOS/L	Sieving	QICPIC/RODOS/L	Sieving	QICPIC/RODOS/L	Sieving	QICPIC/RODOS/L
<25	30.6	13.89	13.84	24.75	37.7	8.1	39.4	14.14
25–40	17.38	13.32	10.91	20.03	10.4	7.93		11.94
40–63	30.18	20.19	59.71	24.77	17.7	14.27	27.6	20.53
63–80	10.02	18.87	12.53	17.41	10.7	20.93	17.4	24.76
80–100	6.6	8.17	1.61	5.31	11.8	12.57	11.3	10.88
100–125	2.45	7.18	0.8	3.012	10.1	13.11		8.49
125–140	2.77	5.97	0.6	2.708	1.6	11.85	4.25	5.11
>140	–	12.41	–	2.01	–	11.24		4.15

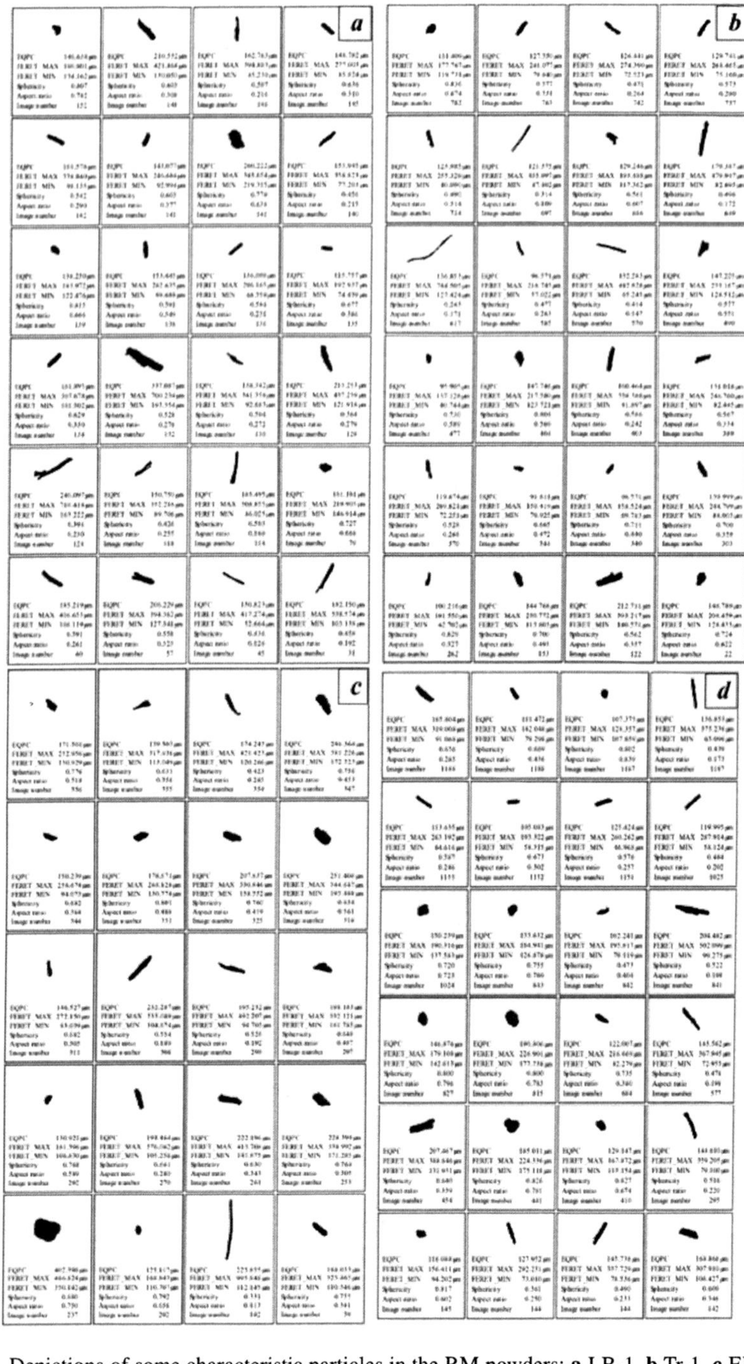

Fig. 2 Depictions of some characteristic particles in the RM powders: **a** LB-1, **b** Tr-1, **c** EK-1, **d** Siberian pine needles (Sympatec QICPIC & RODOS/L analyzer, Germany)

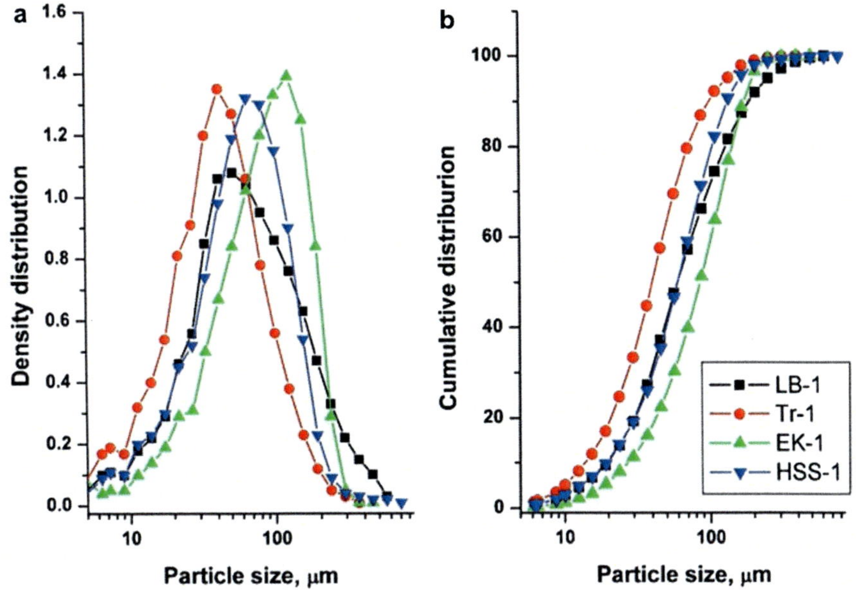

Fig. 3 Integral and differential particle size distributions in the RM powders

Substance Homogeneity and Representative Analytical Weight

The masses of analytical samples required for different methods vary substantially, depending on the functioning principle of each method and the quantity of a test element homogeneously distributed in the sample under study. A large number of modern methods operate on analytical weights from 0.05 to 0.1 g. The result of determining an element will be reliable if the mass of a representative sample of this element in the tested RM is equal to or less than an analytical sample, e.g., 0.1 g. Otherwise, the measurement error increases unpredictably. Therefore, the assessment of particle size distribution and the degree of homogeneity of the distribution of elements in an RM is a mandatory procedure when establishing the certified content and its uncertainty (error) [7].

Direct methods for the analysis of solids without preliminary decomposition of the matrix are more sensitive to the heterogeneity of element distributions in a substance than analytical methods using sample solutions [11, 12]. For the methods with preliminary destruction of a substance, such as ICP-AES or AAS, the analyzed mass of a plant-based RM is often limited by the volumes of containers in which the samples are transferred into solution. However, the decomposition of powder samples greater than 0.10 g usually results in a homogeneous solution from which almost identical in composition aliquots can be taken for analysis, which ensures high accuracy and repeatability of measurements [12]. For non-destructive methods, e.g. XRF or INAA, such an estimate of a representative sample may be underestimated. Therefore, the experimental estimate of the homogeneity for masses of 0.1 and 0.5 g was calculated from the data of ICP-AES and XRF, respectively. Table 2 shows the results of evaluating the homogeneity of element distributions in the HSS-1 RM by these analytical methods. It was experimentally shown that an analytical weight of 0.50 g will be uniquely representative for all the elements and when using any analytical methods (techniques). A representative weight of 0.1 g can be used for determining

Table 2 Assessment of the homogeneity of the HSS-1 material

Analyte	ICP-AES					XRF				
	m = 0.1 g					m = 0.5 g				
	I	S_H, %	V_H, %	γ_H	$w_Д \geq$ (g)	I	S_H, %	V_H, %	γ_H	$w_Д \geq$ (g)
Si	0.14	0.031	0.22	<	0.100	0.61	0.0034	0.006	<	0.5
Al	0.015	0.00	0.13	<	0.100	0.25	0.0008	0.003	<	0.5
Mg	0.12	0.02	0.13	<	0.100	1.2	0.0056	0.005	<	0.5
Ca	0.31	0.05	0.16	<	0.100	31	0.099	0.003	<	0.5
Fe	0.036	0.01	0.15	<	0.100	21	0.067	0.003	<	0.5
Na	0.006	0.0019	0.31	<	0.100	Not defined				
K	0.39	0.02	0.044	<	0.100	31	0.101	0.003	<	0.5
S	0.10	0.0021	0.020	<	0.100	4.6	0.0069	0.001	<	0.5
Ba	579	32	0.056	<	0.100	Not defined				
Sr	7.2	1.37	0.19	<	0.100	1.8	0.012	0.007	<	0.5
Li	0.28	0.02	0.054	<	0.100	Not defined				
P	0.15	0.01	0.048	<	0.100	1.6	0.0057	0.004	<	0.5
B	23	1.42	0.060	<	0.100	Not defined				
Mn	175	20	0.11	>	0.270	4.9	0.019	0.004	<	0.5
Ti	8.5	1.02	0.12	<	0.100	0.88	0.029	0.032	<	0.5
Ni	1.9	0.09	0.045	<	0.100	0.25	0.0067	0.027	<	0.5
Co	20	1.47	0.074	<	0.100	Not defined				
V	0.26	0.02	0.076	<	0.100	Not defined				
Cr	3.0	0.35	0.12	<	0.100	0.16	0.0004	0.002	<	0.5
Sn	0.22	0.02	0.092	<	0.100	Not defined				
Pb	0.4	0.06	0.14	<	0.100	Not defined				
Cu	4.2	0.47	0.11	<	0.100	1.4	0.013	0.009	<	0.5
Zn	39	2.6	0.067	<	0.100	5.4	0.032	0.006	<	0.5
Cd	0.051	0.00	0.074	<	0.100	Not defined				
Se	0.48	0.09	0.20	<	0.100	Not defined				

Note I—the measured intensity of the analytical line of the element; S_H and V_H—heterogeneity error and coefficient of variation of the element; γ_H is an indicator of fulfilling the criterion of a "negligible error", $w_Д$—the minimum representative mass required for a specific determination of the analyte content [13, 14]

Table 3 Major elemental composition of the RMs in the "Plants" cluster

Sample	Mass fraction, %				
	C_{total}	N	O	H	Σ
LB-1	48.0	1.7	40.0	6.4	96.1
Tr-1	42.4	2.2	47.7	6.2	98.5
EK-1	35.5	2.3	52.1	5.4	95.3
HSS-1	49.8	1.4	40.4	6.6	98.2

all tested elements except manganese. When using destructive methods, if an analytical sample will be less than 0.270 g, the heterogeneity of substance will increase the error (uncertainty) in determining manganese. Similar calculations of the representative sample, according to the ICP-AES data for GSO LB-1, Tr-1, and EK-1, showed a possible reduction in the quantity of an analytical sample to 0.10 g for all elements, except iron in LB-1 and calcium in EK-1, if an analytical method provides for the transfer of substance into the solution.

Elemental Composition of Plant RMs

The certified values and the errors of the certified values in plant matrix RMs were established using the method of an interlaboratory comparison. Due to the high content of organic compounds, the contents of oxygen, carbon, hydrogen, and nitrogen were characterized in each sample, and their total amount was indicated (Table 3). When performing the analysis, it was recommended to operate with the representative masses.

Thirty-eight analytical laboratories (24 from Russia and 14 from other countries) took part in the certification studies of LB-1 and performed measurements using 13 methods based on different physical and chemical principles. The interlaboratory certification of Tr-1 was carried out by results of 11 different methods from 36 laboratories (23 in Russia and 13 in other countries). The interlaboratory certification of EK-1 was fulfilled by the results of 10 methods from 23 laboratories (12 in Russia and 11 in other countries). The results of various elements for certificating HSS-1 were presented by 3 to 20 laboratories from Russia and other countries, and obtained using 10 methods. Most of the results were obtained using XRF, AES, and ICP-MS. To assess the stability of the HSS-1 material, only three methods were used: ICP-AES, ICP-MS, and XRF. The error associated with stability (for 20 years) was not taken into account when calculating the uncertainty of RM certification. Calculations of the average contents of elements were carried out in accordance with GOST 8.532–2002 and ISO Guide 35:2015. The established certified characteristics of the content of analytes (A) correspond to the following requirements of GOST 8.532–2002: N—the number of results in the sample must be at least 10; Δ_A—the error of the certified value must be less than the permissible error established in the terms of reference in accordance with the requirements of RMG 76 and OST of the Ministry of Natural Resources and Environment [15–17]. However, a number of elements could not be reliably estimated due to their low contents and insufficient number of results; therefore, these elements were assigned only informational values (Table 4). Thus, the contents of 64–66 elements were characterized in the matrix RMs included in the "Plants" cluster.

The reliability of certificating of the element contents in LB-1, Tr-1, and EK-1 was confirmed by four GWB07602-07,605 GSV-1 $\div$ 4 CRMs of compositions of leaves and branches of the bushes, leaves of poplar and tea. These CRMs were developed to ensure consistency of the analytical data obtained with different analytical methods when fulfilling the "National-Scale Geochemical Mapping Projects in China" [18]. Previously, we described a multidimensional approach to estimating the degree of consistency between certified RMs [19], which relies on the "stem with leaves" method recommended by J. Tukey [20] for performing a multidimensional statistical data analysis. Our study demonstrated the consistency of Russian and Chinese RMs. Currently, three RMs from the "Plant" cluster are used for validating the procedures and quality control of analyzing plant objects using different methods [21, 22].

Conclusion

The "Plants" cluster in the collection of the IGC SB RAS is represented by four RMs: certified RM of the composition of a birch leaf (LB-1) GSO 8923–2007/COOMET RM 0067–2008-RU (*Birch leaf*—LB-1), certified RM of the composition of a grass mixture (Tr-1) GSO 8922–2007/COOMET RM 0066–2008-RU (*Mixture of meadow herbs*—Tr-1), certified RM of the composition of a Elodea Canadensis (EK-1) GSO 8921–2007/COOMET RM 0065–2009-RU (*Canadian pond weed*—EK-1), RM of the composition of a Siberian pine needle (HSS-1) (*Pinus Sylvestris*—HSS-1). These plants are sensitive indicators of pollution of natural terrestrial and aquatic ecosystems. The RMs were collected across areas with the minimum human impact.

The characteristics of these RMs, such as particle size distribution (shape, size, and particle size distribution), homogeneity and minimum representative sample weight, as well as the stability of powders under natural aging conditions, were studied in accordance with the Russian and international requirements using modern instruments and methods of chemical analysis. The elemental compositions of plant-matrix RMs were estimated in terms of more than 60 elements, among which from 23 to 41 were certified.

The developed plant RMs are intended for performing chemical measurements during the validation of existing and development of new methods of analysis, quality control of measurement results and assessment of their traceability, as well as professional tests of geological, agricultural, and pharmaceutical laboratories.

Thus, the use of plant-matrix RMs included in the "Plants" cluster of the RM collection of the IGC SB RAS ensures the accuracy of determining a wide range of

Table 4 Contents of inorganic elements in the plant RMs under study

Element	MU	LB-1		Tr-1		EK-1		HSS-1	
		A	± Δ	A	± Δ	A	± Δ	A	± Δ
Al	%	0.083	0.01	0.037	0.005	0.099	0.012	0.019	0.003
Ca	%	1.6	0.09	0.67	0.03	2.8	0.17	0.42	0.02
Cl	%	*0.045*	*0.009*	0.36	0.04	*0.52*	*0.06*	*0.037*	–
Fe	%	0.073	0.007	0.097	0.005	0.26	0.01	0.047	0.006
K	%	0.71	0.04	1.38	0.03	3.22	0.16	0.46	0.03
Mg	%	0.44	0.03	0.24	0.02	0.32	0.02	0.12	0.02
Na	%	0.018	0.003	0.075	0.006	0.68	0.05	0.0044	0.0004
P	%	0.154	0.006	0.22	0.01	0.24	0.03	0.14	0.02
S	%	0.10	0.02	0.18	0.02	0.34	0.05	0.091	0.010
Si	%	0.4	0.07	0.55	0.04	*1.1*	*0.2*	0.11	0.04
Ag	mg/kg					*0.017*	–	*0.024*	–
As	mg/kg	0.23	0.03	0.16	0.02	0.76	0.12	*0.2*	–
B	mg/kg	*0.005*	*0.001*	11.2	1.2	*33*	*10*	13	3
Ba	mg/kg	0.023	0.002	16.1	1.2	78	7	4.8	0.4
Be	mg/kg	*0.05*	–	*0.014*	–	*0.07*	–	*0.01*	–
Bi	mg/kg					*0.023*	–	*0.0045*	–
Br	mg/kg	3.2	0.4	9	1	32.6	2	*1.2*	–
Cd	mg/kg	0.16	0.03	*0.051*	*0.011*	*0.1*	*0.02*	*0.036*	–
Ce	mg/kg	1.5	0.12	0.5	0.05	3.4	0.3	0.17	0.01
Co	mg/kg	0.79	0.06	0.22	0.02	1.5	0.1	0.17	0.02
Cr	mg/kg	4.3	0.7	5.5	0.4	5.1	0.5	3.6	0.3
Cs	mg/kg	0.085	0.008	0.058	0.008	0.108	0.008	0.019	0.003
Cu	mg/kg	7.3	0.6	6.3	0.6	11.2	0.4	3.8	0.2
Dy	mg/kg	*0.12*	*0.01*	*0.04*	–	*0.36*	*0.13*	*0.013*	–
Er	mg/kg	*0.07*	*0.01*	*0.017*	–	*0.13*	*0.02*	*0.0071*	–
Eu	mg/kg	0.026	0.005	0.0095	0.0014	0.047	0.008	*0.0029*	–
Ga	mg/kg	0.48	0.08	0.16	0.02	*0.4*	–	*0.085*	–
Gd	mg/kg	*0.15*	*0.04*	*0.05*	–	*0.35*	*0.08*	*0.016*	–
Ge	mg/kg	*0.09*	*0.02*	*0.04*	–	*0.07*	–	*0.013*	–
Hf	mg/kg	*0.06*	*0.01*	*0.04*	*0.02*	*0.08*	–	*0.01*	–
Hg	mg/kg	0.037	0.006	*0.015*	–	*0.03*	–	*0.018*	–
Ho	mg/kg	*0.026*	*0.003*	*0.007*	–	*0.047*	*0.008*	*0.0025*	–
La	mg/kg	0.82	0.09	0.26	0.09	2.05	0.14	0.085	0.012
Li	mg/kg	*0.73*	*0.09*	*0.55*	*0.09*	1.44	0.18	*0.21*	*0.02*
Lu	mg/kg	0.011	0.001	*0.0029*	–	0.019	0.003	*0.001*	–

(continued)

Table 4 (continued)

Element	MU	LB-1		Tr-1		EK-1		HSS-1	
		A	± Δ	A	± Δ	A	± Δ	A	± Δ
Mn	mg/kg	930	70	50.9	2.1	520	30	215	12
Mo	mg/kg	*0.16*	*0.06*	0.25	0.02	*1.2*	–	0.12	0.02
Nb	mg/kg	*0.2*	*0.03*	*0.082*	–			*0.037*	–
Nd	mg/kg	0.69	0.06	0.22	0.02	1.59	0.17	*0.072*	*0.013*
Ni	mg/kg	5.8	0.8	3.2	0.3	3.7	0.4	2	0.1
Pb	mg/kg	3.7	0.5	0.42	0.06	*1.1*	*0.1*	0.38	0.05
Pr	mg/kg	*0.19*	*0.02*	0.06	0.009	*0.42*	–	*0.021*	–
Rb	mg/kg	13.7	0.9	15.7	0.4	3.5	0.3	2.3	0.2
Sb	mg/kg	0.057	0.011	*0.019*	*0.005*	*0.08*	*0.02*	0.041	0.007
Sc	mg/kg	0.3	0.04	0.082	0.008	0.38	0.02	0.033	0.004
Se	mg/kg	*0.5*	–	*0.0004*	–	*0.3*	–	*0.04*	–
Sm	mg/kg	0.132	0.015	0.041	0.003	0.31	0.03	*0.015*	–
Sn	mg/kg	*0.19*	*0.09*	*0.09*	–	*0.12*	–	0.15	0.02
Sr	mg/kg	72	7	28	0.9	174	9	11	1
Ta	mg/kg	*0.02*	–	*0.008*	–	*0.017*	*0.004*	*0.012*	–
Tb	mg/kg	0.022	0.003	*0.01*	–	0.041	0.005	*0.002*	–
Th	mg/kg	0.22	0.03	0.055	0.005	0.4	0.03	*0.025*	–
Ti	mg/kg	59	12	33.3	5.3	77	14	11	3
Tl	mg/kg	*0.023*	*0.005*	*0.011*	*0.002*	*0.02*	–	*0.0076*	–
Tm	mg/kg	*0.011*	*0.003*	*0.003*	–	*0.021*	*0.007*	*0.001*	–
U	mg/kg	0.082	0.012	*0.017*	*0.002*	1.4	0.1	*0.093*	–
V	mg/kg	2.1	0.4	0.61	0.09	3.8	0.4	0.27	0.04
W	mg/kg	0.3	0.03	*0.08*	–			*0.0556*	–
Y	mg/kg	0.69	0.06	0.16	0.02	*1.3*	–	0.067	0.005
Yb	mg/kg	0.074	0.007	0.018	0.002	0.074	0.006	0.006	0.002
Zn	mg/kg	94	6	23.6	1.1	20.6	1.4	45	3
Zr	mg/kg	*5.5*	*1.6*	*0.9*	–	*2.6*	–	0.46	0.07

Note A—a certified RM value. Empty cells—the content of the element is uncharacterized; italics marks recommended and informational characteristics

elements by different analytical methods. International plant RMs of other manufacturers are inaccessible for testing laboratories in Russia and, according to the national legislation, cannot be applied without additional administrative procedures.

Acknowledgements The research was conducted within the framework of the state assignment for Projects IX.127.1.4. "Ecological and geochemical transformations of ecosystems in Eastern Siberia under the influence of natural and technogenic factors" No. 0350-2019-0005, No. 0284-2021-0005 "Development of methods for studying the chemical composition and structural state of natural and technogenic environments in the earth sciences" and the Agreement on cooperation between IGC SB RAS and Ltd. "Sympatec".

Author Contributions E. V. Shabanova: experimental research, data analysis, data validation, manuscript draft preparation, manuscript revision and editing.

I. E. Vasil'eva: research supervision, research conceptualization, data curation, data validation, provision of study materials, funding acquisition, manuscript revision and editing.

D. S. Tausenev: data curation, funding acquisition, manuscript revision and editing.

S. Scherbarth: performing experimental research.

U. Pierau: performing experimental research.

Conflict of Interest The article was prepared on the basis of a report presented at the IV International Scientific Conference "Reference Materials in Measurement and Technology" (St. Petersburg, December 1–3, 2020). The article was admitted for publication after the abstract was revised, the article was formalized and the review procedure was carried out.

The version in the Russian language is published in the journal "Measurement Standards. Reference Materials" 2021;17(3):45–61. https://doi.org/10.20915/2687-0886-2021-17-3-45-61.

References

1. Catalogue of certified reference materials of natural and man-made media compositions from A.P. Vinogradov Institute of Geochemisty SB RAS. http://www.igc.irk.ru/images/Innovation/Standartsbr/CATALOGUE_OF_CRMs_IGC_SB_RAS_-2017.pdf
2. Shabanova EV, Vasil'eva IE, Zak AA, Mustafina OK (2018) A new multielement reference material for composition of pine needles. In: *Proceedings Third International Scientific Conference "Reference Materials in Measurement and Technology"*. Ekaterinburg, 11–14 September 2018. UNIIM Publ., Ekaterinburg, Vol. En, pp 146–147
3. ISO Guide 33:2015 (2015) Reference materials. Good practice in using reference materials. BSI, Geneva
4. GOST 8.531–2002 (2003) Reference materials of composition of solid and disperse materials. Ways of homogeneity assessment. Standartinform, Moscow (In Russ)
5. GOST 8.532–2002 (2002) Certified reference materials of composition of substances and materials. Interlaboratory metrological certification. Content and order of works. Standartinform, Moscow (In Russ)
6. GOST R 8.694–2010 (2012) Standard reference materials (substances). General statistical principles of determination of metrological characteristics (MOD ISO Guide 35:2006. Reference materials—General statistical principles of certification). Standartinform, Moscow (In Russ)
7. GOST ISO Guide 35–2015 (2016) Reference materials—General and statistical principles of certification. Standartinform, Moscow (In Russ)
8. Olivares IRB, Souza GB, Nogueira ARA, Toledo GTK, Marcki DC (2018) Trends in developments of certified reference materials for chemical analysis—Focus on food, water, soil, and sediment matrices. Trends Anal Chem 100(3):53–64. https://doi.org/10.1016/j.trac.2017.12.013
9. GOST ISO/IEC 17025–2019 (2020) General requirements for the competence of testing and calibration laboratories. Moscow: Standartinform (In Russ.)
10. R 50.2.031–2003 (2004) Certified reference materials of composition of substances and materials. Ways of stability characteristics assessment. Izdatelstvo Standartov, Moscow (In Russ)
11. Gunicheva TN, Vasileva IE (2012) Studying distribution of some elements in certified reference material of composition of Baikal perch muscle tissues BOk-2 by X-ray fluorescence method. Methods Objects Chem Anal 7(3):132–137 (In Russ)

12. Rossbach M, Zeiller E (2003) Assessment of element-specific homogeneity in reference materials using microanalytical techniques. Anal Bioanal Chem 377(2):334–339. https://doi.org/10.1007/s00216-003-2043-6
13. Khitrov VG, Belousov GE, Bozhevolnova NA, Galuzina AI, Dobronichenko VV, Zemtsova LS, Kopylova LF, Semenov BP (1985) Reliability of rock analysis (facts, problems, solutions). Nauka Publ., Moscow, p 303 (In Russ)
14. Ingamells CO, Switzer P (1973) A proposed sampling constant for use in geochemical analysis. Talanta 20(6):547–568. https://doi.org/10.1016/0039-9140(73)80135-3
15. RMG 76–2014 (2014) Internal control of quantitative chemical analysis result's accuracy. Standartinform, Moscow (In Russ)
16. MPR&E RF OST 41–08–265–2004 (2004) Statistical control of the accuracy (correctness and precision) of the results of quantitative chemical analysis. VIMS, Moscow (In Russ)
17. MPR&E RF OST 41–08–212–2010 (2004) Standards of error in determining the chemical composition of mineral raw materials and classification of methods of analysis according to the accuracy of the results. VIMS, Moscow (In Russ)
18. Yan M, Cheng Z (2007) Study and application of geochemical reference materials in the Institute of Geophysical and Geochemical Exploration (IGGE), China. Geostand Geoanal Res 31(4):301–309. https://doi.org/10.1111/j.1751-908X.2007.00133.x
19. Vasilyeva IE, Shabanova EV, Susloparova VE, Manokhina SN (2014) Assessing extent-of-equivalence of Chinese and Russian certified reference materials of plants using data of mass spectrometry with inductively coupled plasma. Measur Stand Refer Mater 3:24–32 (In Russ)
20. Tukey JW (1977) Exploratory data analysis. Exploring Data Analysis. Pearson, 712 p.
21. Vasil'eva IE, Shabanova EV (2019) Determination of trace elements in plants using the direct current arc atomic-emission spectrometry. Analy Control 23(3):298–313. (In Russ.). http://dx.doi.org/https://doi.org/10.15826/analitika.2019.23.3.011
22. Zaksas NP (2019) Solid sampling in analysis of various plants using two-jet plasma atomic emission spectrometry. Appl Spectrosc 73(8):870–878. https://doi.org/10.1177/0003702819845935

Elena V. Shabanova Dr. Sci. (Phys.-Math.), Senior Researcher, the head of group of atomic emission analysis methods and reference materials, A. P. Vinogradov Institute of Geochemistry, Siberian Branch of the Russian Academy of Sciences (IGC SB RAS).

Irina E. Vasil'eva Dr. Sci. (Eng.), Chief Researcher of the group of atomic emission analysis methods and reference materials, A. P. Vinogradov Institute of Geochemistry, Siberian Branch of the Russian Academy of Sciences (IGC SB RAS).

Dmitrii S. Tausenev Director, Sympatec Representative Office in the Russian Federation.

Stefanie Scherbarth Head of the laboratory, Sympatec GmbH.

Uwe Pierau Head of the laboratory, Sympatec GmbH.

Development and Testing of a Multi-Element Reference Material for Methods Based on Inductively-Coupled Plasma

Tatiana P. Stolboushkina and Aleksei A. Stakheev

Abstract Accurate calibration of the spectrometer output signal in terms of the content of elements under measurement is of great importance for the metrological assurance of high-precision inductively coupled plasma mass spectrometry (ICP-MS) and inductively coupled plasma optical emission spectrometry (ICP-OES). This paper aims to establish the certified values of a reference material for a multi-element solution of metals for use in measurements based on inductively coupled plasma spectrometry (ICP-CRM Multi 1). ICP-CRM Multi 1 is a solution with the certified values of the mass fraction of metals: barium, cadmium, cobalt, lithium, lead, and zinc. The solution was packed in high-density polyethylene bottles with a capacity of 4, 8, 15, 30, 60, and 125 cm^3. The certified values of the mass fraction of metals in the solution was established by the gravimetric method of preparation and confirmed by the State Primary Standard of Unit of Mass Fraction and Unit of Mass (Molar) Concentration of Inorganic Components in Aqueous Solutions Based on Gravimetric and Spectral Methods GET 217–2018. The permissible certified values of the mass fraction of metals in the developed ICP-CRM are shown to range from 900 mg/kg to 1100 mg/kg. The authors have launched a study into the long-term stability of ICP-CRM Multi 1 with the purpose of establishing its expiration date. It is assumed that the expanded uncertainty of measurements of the certified value of the mass fraction of metals in the solution of ICP-CRM Multi 1 will not exceed 0.5%. ICP-CRM Multi 1 can be used for ensuring the metrological traceability of measurements to GET 217–2018 in inorganic analysis using ICP-MS and ICP-OES. The developed solution will also allow one of the main advantages of these methods to be applied in routine analysis, namely the ability to quickly and simultaneously measure several elements in samples.

T. P. Stolboushkina (✉) · A. A. Stakheev
Russian Metrological Institute of Technical Physics and Radio Engineering (VNIIFTRI), Moscow region, Mendeleevo, Russia
e-mail: stolboushkina@vniiftri.ru; mera@vniiftri.ru

A. A. Stakheev
e-mail: stakheev@vniiftri.ru

S. V. Medvedevskikh et al. (eds.), *Reference Materials in Measurement and Technology*,
https://doi.org/10.1007/978-3-031-06285-8_14

Keywords Inductively coupled plasma mass spectrometry · Inductively coupled plasma optical emission spectrometry · Aqueous solution · Reference material · Water analysis · Inorganic component

Introduction

Quality control of industrial products and monitoring of the state of environmental objects are important for both individual enterprises and the state as a whole. Various analytical methods can be used to analyze the chemical composition of substances and materials. The selection of the most appropriate method for the intended purpose is carried out depending on the number of elements to be measured and the number of samples to be analyzed.

As highly rapid and sensitive methods for qualitative and quantitative analysis of elements in various materials and substances, inductively coupled plasma optical emission spectroscopy (ICP-OES) and inductively coupled plasma mass spectrometry (ICP-MS) meet the performance requirements of many laboratories. These methods are increasingly being used in diverse fields of science and industry, including the food industry to control the content of various components in food and drinks [1–3]; environmental protection to analyze the metal content of various environmental objects [4–6]; forensic analysis to identify micro- and toxic elements in biomaterials (hair, nails, epithelial tissue, blood, urine, muscles, etc.) [7–9], etc. [10–12].

The establishment of a calibration dependence of the output signal of spectrometers on the content of elements under analysis is a necessary step for carrying out measurements using ICP-OES and ICP-MS. Such calibration is only appropriate when reliable reference materials (RMs) are used. RMs certified in the Russian Federation (CRM)—GSO—ensure the metrological traceability of measurements in all laboratories of the country [13, 14]. The calibration of spectrometers based on inductively coupled plasma is carried out using CRMs, which are solutions with the certified value of the mass fraction or mass concentration of a particular element of Mendeleev's Periodic Table.

The methods of ICP-MS and ICP-OES allow several elements to be measured quickly and simultaneously. This goal can only be achieved when using multi-element RMs, since their use simplifies and accelerates the preparation of a series of working and calibration solutions.

In this research, we aimed to develop a CRM of a multi-element solution of metals (hereinafter ICP-CRM Multi 1), which would enable the calibration and measurement of several elements simultaneously. The CRM under development should be traceable to the State Primary Standard of Unit of Mass Fraction and Unit of Mass (Molar) Concentration of Inorganic Components in Aqueous Solutions Based on Gravimetric and Spectral Methods GET 217–2018 [15].

Materials and Methods

In order to develop a CRM of a multi-element solution of metals, the following metrological characteristics required normalization: the value of a CRM certified characteristic; error and/or uncertainty of a CRM certified value; the expiration date of a CRM.

The values of CRM certified characteristics were normalized by establishing the interval across which the certified values of any RM of a given type must vary. The mass fraction of metals in the developed CRM should range from 900 mg/kg to 1100 mg/kg inclusive; the mass concentration of a metal should range from 900 mg/dm^3 to 1100 mg/dm^3 inclusive.

The selection of a manufacturing technology was conducted by analyzing possible procedures for preparing an ICP-CRM and approaches to establishing the ICP-CRM certified values. As a result, we decided to certify the developed ICP-CRM according to the gravimetric method of preparation, since the gravimetric preparation of RMs from high-purity starting materials is the most accurate method for reproducing and transferring the units of concentration of metals in a solution. Certification by the gravimetric preparation procedure is an approach to certifying an RM based on the known or specially investigated characteristics and quantitative ratios of the precursor components used for RM preparation by mixing them to obtain the calculated values of the RM metrological characteristics.

The material chosen for preparing the ICP-CRM is a solution of metals or their compounds in dilute nitric acid. The mass fraction of the main component in the starting materials and the solvent was established using GET 217–2018. The ICP-CRM was produced using the VNIIFTRI equipment. In order to carry out preliminary experiments, pure metals were chosen as a starting material for preparing a solution with a certified value of the mass fraction of zinc, cobalt, and cadmium; metal salts were chosen for a solution with a certified value of the mass fraction of lithium, barium, and lead. Bottles made of darkened high-density polyethylene with various capacities (30, 60, and 125 cm^3) were selected as containers. They had been previously used in the development of single-element CRMs [16]. The solvent was 5% nitric acid by volume, which ensured complete dissolution of the starting material and stability of the ICP-CRM in accordance with the preparation procedure. Particular attention was paid to the selection and pre-cleaning of polymer ware intended for both laboratory research and further use as a container for the ICP-CRM.

In order to assess the purity of the starting materials, the "100% minus the amount of impurities" method was chosen. Thus, all detected impurities were subtracted from 100%, and elements with a concentration below the detection limit were taken into account by subtracting half of their detection limit from 100%. The mass fraction of the main component (ω) was calculated by Eq. (1):

$$\omega = 100\% - \sum_i w_i - \sum_j \frac{LOD_j}{2} \qquad (1)$$

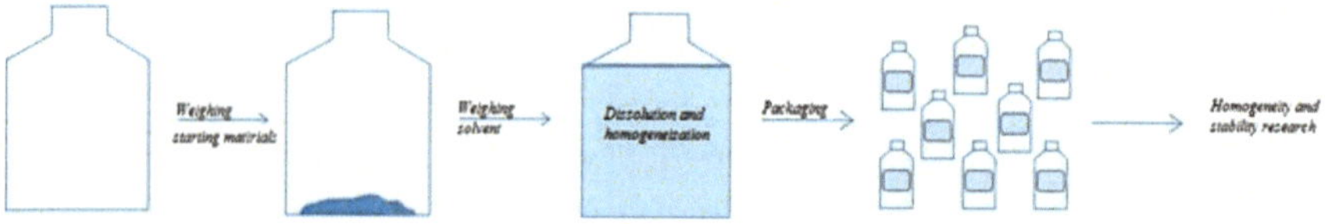

Fig. 1 A scheme of the ICP-CRM Multi 1 manufacturing technology

where w_i is the mass fraction of detected impurities, %;

LOD_j is the detection limit of undetected impurities, %.

The impurity composition of the solvents after purification was assessed using GET 217–2018.

The mass fraction of a component in the ICP-CRM was estimated by equations [17]:

$$\overline{A_{m_i}} = \sum_{i=1}^{N} w_j \cdot a_{m_{ij}} \tag{2}$$

$$w_j = \frac{m_j}{\sum_{j=1}^{N} m_j} \tag{3}$$

where $\overline{A_{m_i}}$ is the weighted average estimate of the content of the *i*-th component in the mixture;

$a_{m_{ij}}$ amis the content of the *i*-th component in the *j*-th component in the mixture;

w_j are weight coefficients;

m_j is the mass of the *j*-th component in the mixture.

A scheme of the ICP-CRM Multi 1 manufacturing technology from carrier materials is shown in Fig. 1. The mass fraction of the components of the carrier materials was preliminarily established according to the "100% minus the $\sum$amount of impurities" method using GET 217–2018.

The traceability of the certified values of the ICP-CRM to the measurement unit was carried out by establishing the value of the certified characteristics in the starting materials using GET 217–2018.

Results and Discussion

The developed ICP-CRM Multi 1 was packed in bottles with a capacity of 30, 60, and 125 cm^3 (Fig. 2).

The reliability of the certified values of the mass fraction of metals in the developed ICP-CRM was repeatedly measured using GET 217–2018. The measurement results of the mass fraction of metals in the ICP-CRM obtained using both the gravimetric

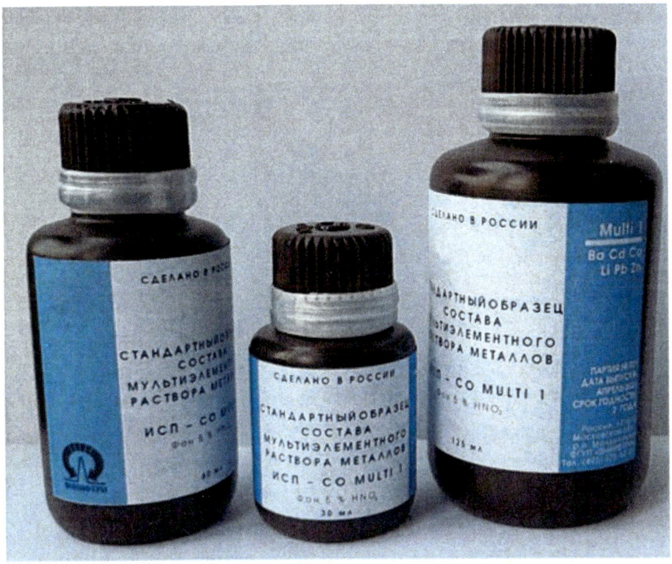

Fig. 2 The ICP-CRM Multi 1

sample preparation and the measurement method using GET 217–2018 are in good agreement within the limits of their uncertainties, as well as with the results of comparative measurements carried out for the developed ICP-CRM and CRMs of other manufactures. The certified value of the developed ICP-CRM is taken to be that obtained by the gravimetric method of preparation.

The measurement uncertainty of the certified value of the mass fraction of a component in the solution was calculated in accordance with [18]. The uncertainty components for the gravimetric preparation of the solution of the mass fraction of a component are presented in Fig. 3 as a cause-and-effect diagram. The results of the gravimetric preparation and the measurement results using GET 217–2018 are presented in Fig. 4.

The results of estimating the uncertainty due to the method for determining the certified value of the CRM of a multi-element solution of metals according to the gravimetric method of preparation are presented in Tables 1 and 2.

The uncertainty of a certified value includes the uncertainty due to instability and the uncertainty due to homogeneity, in addition to the uncertainty due to the method for determining the certified value. Therefore, future research should investigate the stability and homogeneity of the developed ICP-CRM.

At the current research stage, special attention was paid to the selection of both suitable packaging for storing the developed CRM and its stable composition. However, the mass fraction of any component in a solution is a time function and may change due to possible evaporation losses during storage after unpacking. In this regard, we are planning to conduct additional studies into the long-term stability of the certified value of ICP-CRM Multi 1 and its stability during storage

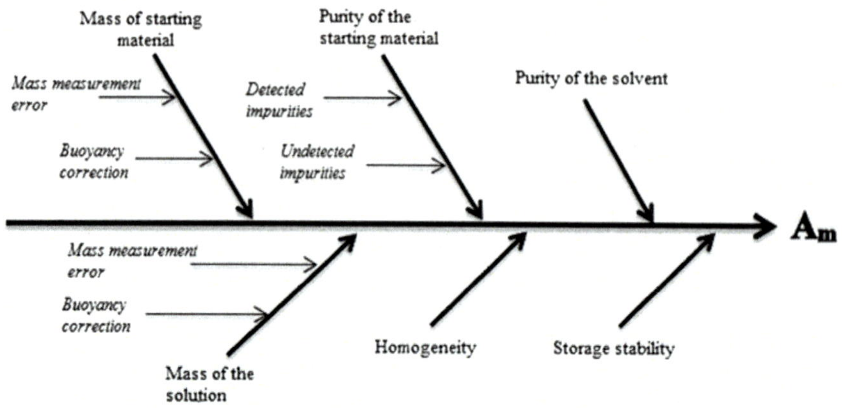

Fig. 3 A cause-and-effect diagram for the error components of a CRM certified value

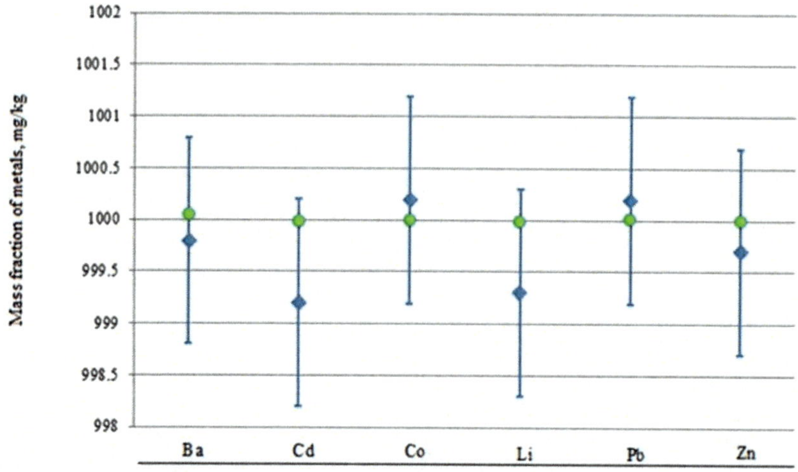

	Ba	Cd	Co	Li	Pb	Zn
Gravimetric sample preparation	1000.05	999.98	999.99	999.98	1000.01	1000.00
GET 217-2018	999.80	999.20	1000.20	999.30	1000.20	999.70

Fig. 4 The results of the gravimetric preparation and the measurement results using GET 217–2018

after unpacking. Provided that positive results are obtained in terms of long-term stability, the expanded measurement uncertainty of the certified value of the metal mass fraction in the CRM aqueous solution will not exceed 0.5%.

Table 1 The uncertainty due to the method for determining the certified value of the CRM of a multi-element solution of metals based on the results of the gravimetric sample preparation

Element	A_m, mg/kg	Standard uncertainty for the method of certification, mg/kg	Relative standard uncertainty for the method of certification, %
Ba	1000.05	1.44	0.145
Cd	999.98	1.75	0.175
Co	999.99	1.45	0.145
Li	999.98	1.59	0.160
Pb	1000.01	1.20	0.120
Zn	1000.00	1.55	0.155

Table 2 The uncertainty budget for gravimetric preparation of the CRM of a multi-element solution of metals on the example of cobalt

Source	Value	Measurement unit	Standard uncertainty	Contribution
Sample weight of barium carrier material	12.9343	g	$289 \cdot 10^{-6}$	$32 \cdot 10^{-6}$
Mass fraction of the certified component (cobalt) in the starting barium carrier material	$32.80 \cdot 10^{-6}$	%	$9.18 \cdot 10^{-6}$	$130 \cdot 10^{-6}$
Sample weight of cadmium carrier material	9.0016	g	$289 \cdot 10^{-6}$	$32 \cdot 10^{-6}$
Mass fraction of the certified component (cobalt) in the starting cadmium carrier material	$11.50 \cdot 10^{-6}$	%	$4.45 \cdot 10^{-6}$	$44 \cdot 10^{-6}$
Sample weight of cobalt carrier material	9.0013	g	$289 \cdot 10^{-6}$	0.032
Mass fraction of the certified component (cobalt) in the starting cobalt carrier material	99.998	%	0.145	1.4
Sample weight of lithium carrier material	95.5431	g	$289 \cdot 10^{-6}$	$32 \cdot 10^{-6}$

(continued)

Table 2 (continued)

Source	Value	Measurement unit	Standard uncertainty	Contribution
Mass fraction of the certified component (cobalt) in the starting lithium carrier material	$36.4 \cdot 10^{-6}$	%	$10.5 \cdot 10^{-6}$	$1.1 \cdot 10^{-3}$
Sample weight of lead carrier material	14.3881	g	$289 \cdot 10^{-6}$	$32 \cdot 10^{-6}$
Mass fraction of the certified component (cobalt) in the starting lead carrier material	$408.0 \cdot 10^{-6}$	%	$60.6 \cdot 10^{-6}$	$970 \cdot 10^{-6}$
Sample weight of zinc carrier material	9.0532	g	$289 \cdot 10^{-6}$	$32 \cdot 10^{-6}$
Mass fraction of the certified component (cobalt) in the starting zinc carrier material	$66.0 \cdot 10^{-6}$	%	$762 \cdot 10^{-9}$	$7.7 \cdot 10^{-6}$
Solvent weight	8851.6	g	0.577	0.064
Mass fraction of the certified component (cobalt) in the solvent	$2.1 \cdot 10^{-6}$	%	$115 \cdot 10^{-9}$	$1.1 \cdot 10^{-3}$
Mass fraction of cobalt in the CRM of a multi-element solution of metals	999.99	mg/g		
Combined standard uncertainty	1.45	mg/g		

Conclusion

In this research, we have developed a CRM of a multi-element solution of metals (ICP-CRM Multi 1), which can be used for ensuring the metrological traceability of measurements in inorganic analysis using ICP-MS and ICP-OES according to GET 217–2018. The developed solution will also allow one of the main advantages of these methods to be applied in routine analysis, namely the ability to quickly and simultaneously measure several elements in materials.

Acknowledgements All measurements were carried out using the equipment of VNIIFTRI.

Author Contributions Stolboushkina T. P.: conducting experiments, collection of experimental data, development of preparation and measurement methods, literature review.

Stakheev A. A.: development of the research and article concept, critical analysis and revision of the text.

Conflict of Interest The article was prepared on the basis of a report presented at the IV International Scientific Conference "Reference Materials in Measurement and Technology" (St. Petersburg, December 1–3, 2020). The article was admitted for publication after the abstract was revised, the article was formalized and the review procedure was carried out.

The version in the Russian language is published in the journal "Measurement Standards. Reference Materials" 2021;17(2):49–57. https://doi.org/10.20915/2687-0886-2021-17-2-49-57.

References

1. Tokalıoğlu Ş, Dokan FK, Köprü S (2019) ICP-MS multi-element analysis for determining the origin by multivariate analysis of red pepper flakes from three different regions of Turkey. LWT- Food Sci Technol 103:301–307. https://doi.org/10.1016/j.lwt.2019.01.015
2. Nguyen-Quang T, Do-Hoang G, Truong-Ngoc M (2021) Multielement analysis of pakchoi (Brassica rapa L. ssp. chinensis) by ICP-MS and their classification according to different small geographical origins. J Analy Methods Chem. https://doi.org/10.1155/2021/8860852
3. Gregorčič SH, Strojnik L, Potočnik D, Vogel-Mikuš K, Jagodic M, Camin F (2020) Can we discover truffle's true identity? Molecules 25(9):2217. https://doi.org/10.3390/molecules25092217
4. Sader JA, Ryan S (2020) Advances in ICP-MS technology and the application of multi-element geochemistry to exploration. Geochem Expl Environ Anal 20(2):167–175. https://doi.org/10.1144/geochem2019-049
5. Carrier-Belleau C, Drolet D, McKindsey CW, Archambault P (2021) Environmental stressors, complex interactions and marine benthic communities' responses. Scient Rep Feb;11(1):4194. https://doi.org/10.1038/s41598-021-83533-1
6. Noerpel M, Pribil M, Rutherford D et al (2020) Lead speciation, bioaccessibility and source attribution in Missouri's Big River watershed. Appl Geochem J Intern Assoc Geochem Cosmochem 123. https://doi.org/10.1016/j.apgeochem.2020.104757
7. Vanini G, Souza MO, Carneiro MTWD, Filgueiras PR, Bruns RE, Romão W (2015) Multivariate optimisation of ICP OES instrumental parameters for Pb/Ba/Sb measurement in gunshot residues. Microchem J 120:58–63. https://doi.org/10.1016/j.microc.2015.01.003
8. Galay EP, Dorogin RV, Temerdashev AZ (2021) Quantification of cobalt and nickel in urine using inductively coupled plasma atomic emission spectroscopy. Heliyon 7(1):e06046. https://doi.org/10.1016/j.heliyon.2021.e06046
9. Wilk A, Romanowski M, Wiszniewska B (2021) Analysis of cadmium, mercury, and lead concentrations in erythrocytes of renal transplant Recipients from Northwestern Poland. Biology 10(1):62. https://doi.org/10.3390/biology10010062
10. Kreitals NM, Watling RJ (2014) Multi-element analysis using inductively coupled plasma mass spectrometry and inductively coupled plasma atomic emission spectroscopy for provenancing of animals at the continental scale. Forensic Sci Intern Synergy 244:116–121. https://doi.org/10.1016/j.forsciint.2014.08.016
11. Haraguchi H et al (1994) Multielement profiling analysis of geochemical and environmental samples by inductively coupled plasma-atomic emission spectroscopy (ICP-AES) and inductively coupled plasma-mass spectrometry (ICP-MS). In: Conference Proceedings. 10th Anniversary. IMTC/94. Advanced Technologies in I & M. 1994 IEEE Instrumentation and Measurement Technolgy Conference (Cat. No.94CH3424–9). Hamamatsu, Japan, pp 1–6. https://doi.org/10.1109/IMTC.1994.351792
12. Michalak I, Chojnacka K (2009) Multielemental analysis of macroalgae from the Baltic sea by ICP-OES to monitor environmental pollution and assess their potential uses. Int J Environ Anal Chem 89(8–12).583–596. https://doi.org/10.1080/03067310802627213
13. De Bièvre P, Dybkaer R, Fajgelj A, Hibbert DB (2011) Metrological traceability of measurement results in chemistry: concepts and implementation (IUPAC Technical Report). Pure Appl Chem 83(10):1873–1935. https://doi.org/10.1351/PAC-REP-07-09-39

14. Fed Federal Law (2008) № 102-FZ of 26/06/2008 "On Ensuring the Uniformity of Measurements" (as amended and supplemented). https://normativ.kontur.ru/document?moduleId=1&documentId=398044 (In Russ)
15. GET 217–2018 National primary standard for the units of mass fraction and mass (molar) concentration of inorganic components in aqueous solutions based on gravimetric and spectral methods. In: Federal information fund for ensuring the uniformity of measurements https://fgis.gost.ru/fundmetrology/registry/12/items/397906 (In Russ)
16. Stakheev AA, Stolboushkina TP (2019) Developing and testing a certified reference material of the lead mass fraction in solutions for measurements using inductively coupled plasma methods. Reference Materials 15(4):25–31. https://doi.org/10.20915/2077-1177-2019-15-4-25-31 (In Russ)
17. Migal PV, Medvedevskikh SV, Firsanov VA (2020) A method for estimating the certified value uncertainty of a multicomponent reference material. In: Medvedevskikh S, Kremleva O, Vasil'eva I, Sobina E (eds) Reference Materials in Measurement and Technology Springer, Cham. pp 175–185. https://doi.org/10.1007/978-3-030-32534-3_17
18. JCGM 100:2008. Evaluation of measurement data—Guide to the expression of uncertainty in measurement (GUM). Joint Committee for Guides in Metrology
19. MI 3560–2016 Recommendation (2016) GSE. Estimation of uncertainty in measurements of the mass fraction of the main component in inorganic substances. UNIIM, Ekaterinburg
20. Migal PV, Gorbunova EM, Sobina EP, Tabatchikova TN (2013) Development and testing of a reference material of nickel (II) mass fraction in solution. Ref Mater 3:39–44 (In Russ)

Tatiana P. Stolboushkina Engineer of the laboratory, the Federal State Unitary Enterprise «Russian metrological institute of technical physics and radio engineering» (FSUE «VNIIFTRI»).

Aleksei A. Stakheev Head of the laboratory, the Federal State Unitary Enterprise «Russian metrological institute of technical physics and radio engineering» (FSUE «VNIIFTRI»).

Plant-Matrix Reference Materials as a Tool for Ensuring the Unity of Chemical Measurements in Geochemistry, Ecology, Agriculture, and Pharmacology

Irina E. Vasil'eva and Elena V. Shabanova

Abstract In order to preserve the environment and its safety for the population, reliable information is needed about the chemical composition of plants and various products on their basis. An increase in cross-border trade advances the demand for traceable results obtained in determining the content of chemical elements, which, along with proteins, fats, carbohydrates, pesticides, moisture, vitamins, etc., affect the quality of human life. The reliability of elemental composition measurements in agricultural and wild plants, as well as various plant products, appears to be an urgent but challenging analytical task. Representing a generally recognized tool for ensuring the uniformity of chemical measurements, reference materials (RMs) are intended for certification (validation) of existing and new methods (techniques) of chemical analysis, certification studies in the development of RMs, as well as the proficiency testing of laboratories. The present article lists the authoritative manufacturers of plant-matrix reference materials with the certified content of chemical elements. The ratio of certified, reference, and quality control plant-matrix materials is estimated. The classification of certified RMs according to the type of plant material used for their food purpose is given. The contribution of different countries to the development of plant-matrix RMs is shown. The selection of plants for the development of new RMs is discussed from two perspectives, namely the nutrition facts system (AOAC INTERNATIONAL) and the paradigm of the "Reference Plant" fingerprint (B. Markert). Based on the generalized studies concerning the development and use of plant-matrix RMs, a list of the most important requirements is compiled with respect to RMs ensuring the reliability and comparability of the results of chemical analysis in the fields of biology, geochemistry, ecology, agriculture, medicine, and interdisciplinary research.

I. E. Vasil'eva (✉) · E. V. Shabanova
A.P. Vinogradov Institute of Geochemistry, Siberian Branch of the Russian Academy of Sciences (IGC SB RAS), 1A Favorsky str, Irkutsk 664033, Russia
e-mail: vasira@igc.irk.ru

E. V. Shabanova
e-mail: shev@igc.irk.ru

S. V. Medvedevskikh et al. (eds.), *Reference Materials in Measurement and Technology*,
https://doi.org/10.1007/978-3-031-06285-8_15

Keywords Ensuring the uniformity of measurements · Plant reference materials · Elemental composition of plants · List of requirements for plant-matrix reference materials

Introduction

As a part of the Earth's biosphere, flora represents one of the main groups of multicellular organisms that appeared on Earth more than 3 billion years ago. Plants contain practically all chemical elements of D.I. Mendeleev's Periodic Table. Currently, about 320 thousand plant species are known [1]. The living matter of plants (over 90%) consists mainly of organic compounds (proteins, carbohydrates, and fats) and water. The latter makes up 70–95 and 5–15% in most vegetative organs of plants and seeds, respectively. Organomineral and mineral components make up a small fraction of living matter. Organic compounds appear to be quite unstable over time due to possible decomposition under the influence of moisture, light, and microorganisms (fermentation, decay). Dry matter of plants contains 90–95% of organic compounds and 5–10% of mineral salts and organoelement compounds, including up to 92 chemical elements (CE). By virtue of biological selectivity for chemical elements, plants are capable of controlling their chemical composition. The role of each chemical element in a plant's life can be assessed by determining its gross content and local forms of presence involved in the processes of transfer, metabolism, and accumulation.

Traditionally, plants are used for nutrition, treatment of diseases, provision of clothing and technical fabrics, housing construction, energy generation, etc. Human activity changes the flora of the Earth. The results of the plant chemical analysis are applied to study the biological diversity of plants and describe the ecological state of territories, as well as to assess the quality of food products, medicinal plant raw materials (MPRM), and preparations based on them, which are established to be safe for human, animal, and poultry health. As an urgent analytical task, reliable and comparable measurements of the elemental composition in agricultural and wild plants, food products, medicinal, cosmetic, and hygienic preparations on a plant basis also become an important factor in preserving the safe environment for human life.

Depending on the tasks of research projects or production requirements, in the analysis of plants, various single and multielement analytical methods (techniques) are used for determining the gross and local concentrations of CE in individual organs and even plant cells without changing the aggregate state of the samples or by transferring plant samples into solution. Until the 70 s of the last century, analytical laboratories prepared and used their own single- and multielement solutions for calibration and verification of titrimetric, colorimetric, and spectrophotometric techniques and instruments widely applied in chemical analysis. With an increase in the amount of analytical work, variations in the macrocomposition of multielement solutions were found to cause a significant effect on the results of the trace element determination. In 1964, in order to provide an agreement in the determinations of trace elements in plants, Professor H.J.M. Bowen from the Agricultural Institute (Great Britain) grew cabbage Marrow Stem Kale (Brassica oleracea) and prepared the first plant-matrix RM from its leaves [2]. This RM became popular in analytical laboratories and the content of 60 chemical elements was certified in it according to the results of inter-laboratory comparisons. As a result, RMs started to be applied to control the variants of dissolving plant-matrix samples, calibrate methods, as well as to take into account matrix effects and spectral overlays. Currently, according to Federal Law FZ-102 [3] and ISO 17025–2019 [4], plant-matrix certified reference materials (CRM), reference materials (RM), and quality control samples (QC) represent a generally recognized tool for ensuring the uniformity of chemical measurements. They are intended for certification (validation) of existing and new methods (techniques) of chemical analysis, certification studies in the development of RMs (i.e., when assessing the average content of analytes and their uncertainty), and proficiency (qualification) testing of laboratories [5–8].

This article focuses on particular plant-matrix RMs with a certified content of individual CEs. Here, the number (quantity), quality, and information accessibility of plant-matrix reference materials from different manufacturers are discussed with regard to the unity of chemical measurements in geochemical, environmental, agricultural and medico-biological projects [9]. The article reviews materials on the development and use of plant RMs [9], as well as generalizes the requirements and criteria for plant RMs used in geoecological research, agriculture, and pharmaceuticals.

Materials and Methods

We collected 270 publications, including international and Russian regulatory documents, where the problems of the development and use of plant-matrix reference materials are considered for the period from the 1960s to the present [9]. Information about the developed RMs was retrived from the websites of RM manufacturers, Russian and international catalogs and electronic databases, such as FSIS "Arshin," COMAR, and "GeoRem".[1] In addition, a search for plant RMs for the study of [9] was carried out on the websites of leading RM manufacturers.[2]

[1] FSIS "Arshin" - Federal Information Fund for Ensuring the Uniformity of Measurements of ROSSTANDART (Russia). Available at: http://fundmetrology.ru.

COMAR - International database of certified reference materials (BAM, Germany). Available at: http://www.comar.bam.de.

GeoRem - Geological and environmental reference materials (Max Planck Institute for Chemistry, Germany). Available at: http://georem.mpch-mainz.gwdg.de.

[2] NIST - National Institute of Standards and Technology, formerly NBS - National Bureau of Standards, USA. Available at. https://www.nist.gov.

NRC - National Research Council, Canada. Available at: https://nrc.canada.ca/en.

JRC - Joint Research Center, Directorate-General of the European Commission (EC).

IAEA - International Atomic Energy Agency. Available at: https://www.iaea.org.

The certificates of plant-matrix RM of the listed manufacturers contain certified, recommended, and informational contents of chemical elements and/or their isotopes. The contents of up to 35–80 certified and recommended elements are provided in the certificates of some CRMs (NIST, NRC, BCR, INCT, BelGIM, UNIIM, IGC SB RAS, etc.) and QC samples (WEPAL-IPE).

Publications consider almost all existing analytical methods for obtaining elemental composition of various objects, including plants and their organs (roots, stems, bark, leaves, flowers, fruits, sap), food and medicinal products based on them (grain, cereals, flour, tea, coffee, cocoa, spices, juices, vegetable and essential oils, tinctures, extracts, vitamins, dietary supplements, etc. [9]. Nevertheless, each branch of knowledge (agrochemistry, botany, biochemistry, biogeochemistry, ecology, pharmacology, etc.) is characterized by traditional sets (complexes) of analytical methods significantly differing from each other. This is explained by the fact that, firstly, the lists of the determined elements vary and, secondly, none of the existing analytical methods can provide for simultaneously collecting complete information even on the gross contents of both macro- and microelements.

Depending on the tasks of research projects or production requirements, the analysis of plants can be performed by all currently existing analytical methods: chemical, physicochemical, and physical; direct and indirect; with and without changing the aggregate state of the samples; single- and multielement; gross and local, etc.

BAM - Federal Institute for Materials Research and Testing, Germany. Available at: https://www.bam.de/Navigation/DE/Home/home.html.

WEPAL-IPE – CII – Comitê Inter-Instituts of the University of Wageningen (Netherlands) for the International Plant-Analytical Exchange (IPE). Available at: https://www.wepal.nl.

INCT - Nuclear Chemistry and Technology Institute, Poland. Available at: https://www.euronuclear.org/dt_team/poland.

BelGIM - Belarusian State Institute of Metrology - National Metrological Institute of the Republic of Belarus. Available at: cess mode: http://belgim.by.

Pryanishnikov Institute of Agrochemistry, formerly TsINAO - Central Institute of Agrochemical Services for Agriculture, Moscow, Russia. Available at: https://www.vniia-pr.ru.

D.I. Mendeleev All-Russian Institute for Metrology (VNIIM), St. Petersburg, Russia. Available at: https://www.vniim.ru/index.html.

UNIIM - a branch of D.I. Mendeleev All-Russian Institute for Metrology (VNIIM), Yekaterinburg, Russia. Available at: https://uniim.ru.

IGC SB RAS - A.P. Vinogradov Institute of Geochemistry, Siberian Branch of the Russian Academy of Sciences, Irkutsk, Russia. Available at: http://www.igc.irk.ru/ru.

Ekolan - Scientific-Production and Analytical Center "Ekolan," Moscow, Russia Available at:: http://www.ecolan.com.ru.

IGGE and NIM - Institute of Geophysical and Geochemical Exploration and National Metrology Institute, China. Available at:: https://en.nim.ac.cn.

NIES and NMIJ - National Institute for Environmental Studies and National Metrology Institute, Japan. Available at:: https://www.nies.go.jp/index-e.html.

KRISS - Korea Research Institute of Standards and Science. Available at:: https://www.kriss.re.kr/eng/main/main.html.

INMETRO National Institute of Metrology, Standardization and Industrial Quality, Brasília. Available at:: http://www.inmetro.gov.br.

ANARL - Australian National Analytical Reference Laboratory. Available at: https://www.nata.com.au/about-nata.

and other manufacturers of certified reference materials.

The evolution of methods for plant analysis is carried out through optimization and improvement of procedures of complete or partial (group) dissolution of plant samples, as well as group (individual) methods of concentration and separation of CE; application of sorption/extraction of CE by new organic reagents, compounds or their mixtures; development of new analytical equipment, automation and computerization in obtaining and processing of measured data. In combination with the improvement of the metrological support of analytical research, this leads to the unification and rapidity of the analysis with improved detection limits and increased accuracy of the results.

For routine production control, analytical procedures should be simple and rapid, having the detection limits below the controlled level of the analyte. In determination of toxic elements, it is especially important for an analytical technique to have a detection limit or a lower limit of the determined content 2–10 times lower than the MPC or permissible daily exposures.

Verification of analytical methods involves the analysis of the corresponding matrix RMs in order to assess the influence of the special features of plant objects on the selection and homogenization of the substance, chemical treatment, and measurement.

Results and Discussion

Despite the wide diversity of plant species [1], the total list of multielement RMs for ensuring the correctness and traceability of the results in determining a wide range of elements by different analytical methods turned out to be very modest and amounted to about 1000 samples [9]. The list of the discussed RMs (Fig. 1) includes certified reference materials (CRMs), reference materials (RMs), and quality control

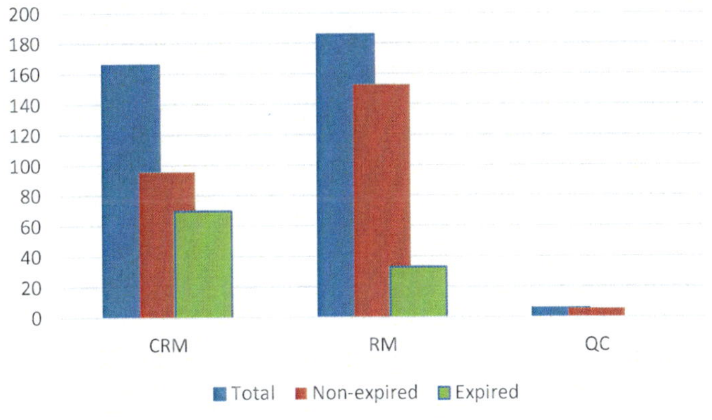

Fig. 1 The number of plant-matrix reference materials by the type of certification (information on open data as of 03/01/2020) [9]

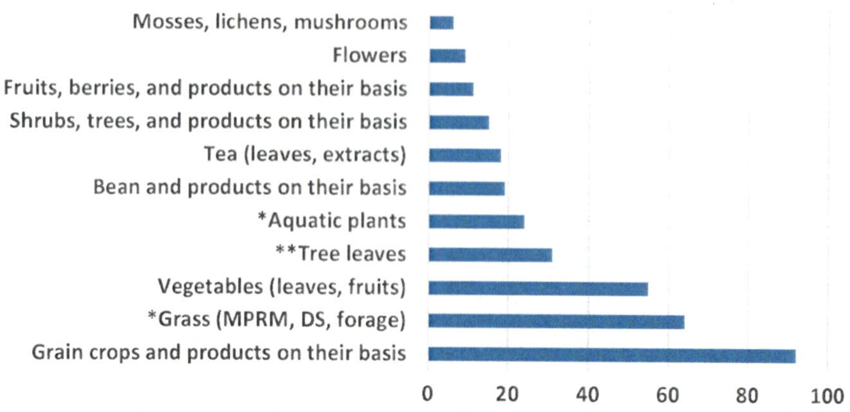

Fig. 2 Classification of reference materials by the type of plant material (information on open data as of 03/01/2020). The belonging of four plant-matrix RM from the IGC SB RAS collection of reference materials to the indicated types is marked with asterisks

samples (QC). The list also included the first developed plant reference materials, which are currently practically exhausted, and RMs with expired shelf life (Fig. 1). The group of non-certified RMs accounts for the largest number of developed RMs, while the minimum number of materials is characteristic of the quality control group. Obviously, RMs are in maximum demand. In this group, the share of RMs with expired shelf life is significantly lower than that of the CRM group, although the total number of CRMs and non-certified RMs differs insignificantly. The main reason for the limited shelf life of RMs involves the instability of the organic matrix over time.

Typically, RMs are prepared in the form of powders, tablets or granules, as well as aqueous or alcoholic extracts and oils. The discussed RMs were divided into groups according to the type of plant material and their food purpose (Fig. 2). The RMs from the listed groups are used for analytical control in agricultural, agrochemical, food, and pharmaceutical laboratories and in the implementation of research projects (geoecological monitoring, conservation of botanical diversity, study of plant nutrition, etc.). The division of RMs by plant species and consumer qualities (Fig. 2) indicates the high nutritional value of agricultural crops for the population of all countries around the world. Four plant RMs (marked with asterisks in the corresponding groups in Fig. 2) were developed at the Institute of Geochemistry SB RAS: CRM of birch leaf LB-1,[3] CRM of meadow herb mixture Tr-1,[4] CRM of Canadian

[3] GSO 8923–2007 Certified reference material for composition of birch (LB-1). Available at: https://fgis.gost.ru/fundmetrology/registry/19/items/391262.

COOMET CRM 0067–2008-RU CRM for composition of birch *Birch leaf* (LB-1). Available at: http://www.coomet.org/en/doc/i2_2008.pdf.

[4] GSO 8922–2007 Certified reference material for composition of meadow herb mixture (Tr-1). Available at: https://fgis.gost.ru/fundmetrology/registry/19/items/391263.

COOMET CRM 0066–2008-RU CRM for composition of mixture *Mixture of meadow herbs* (TR-1). Available at: http://www.coomet.org/en/doc/i2_2008.pdf.

pond weed EK-1,[5] RM of pine needles KhSS-1 (Pinus Sylvestris) [10]. The application of these multielement RMs is aimed at increasing the reliability of analytical data during geoecological monitoring. Each RM indicates the pollution of natural terrestrial and aquatic ecosystems. The selection of an RM form is based on such specific features of plants as the ability to rapidly change the elemental composition under the conditions of environmental pollution, wide growing area, and the possibility of sampling biomass for the preparation of RM in a sufficient amount. The above RMs are certified (recommended) for the content of 41(22), 38(25), 34(30), and 28(12) CE, respectively. Metrological characteristics (average values and their errors) were established by a round robin test according to the results of more than 25 analytical laboratories and 10–13 methods based on different physical and chemical principles applied. Certified RMs with established metrological traceability were used to control the accuracy of the results.

The diagram in Fig. 3, published in [9], demonstrates the proportion of plant-matrix RMs developed in different countries and indicates the high importance of the safety problem of food and medicinal products practically all over the world. In accordance with its geographic location, climatic and national-cultural characteristics, each country has its own list of priority traditionally cultivated crops and food products made on their basis, medicinal plant raw materials and preparations. Cross-border trade promotes the spread of food and spices from wild or cultivated endemic plants such as different varieties of tea, coffee, cocoa beans (Colombia, Kenya, Senegal, Brazil), rice and soybeans (Japan, Korea, China), fruits - bananas, apples, tangerines (Ecuador, Argentina, Morocco, etc.), medicinal plant raw materials, medicines, dietary supplements (India, Tibet, China), etc. The quality and safety of food products and medicinal products depend on the accumulated amount of trace elements and compounds harmful to human health and are ensured by the traceability of the results to the same CRM during analytical control.

The International Association of Official Agricultural Chemists (AOAC INTERNATIONAL) uses an organizational food composition system allowing for the comparison of different food products on a single basis and examining their composition by different analytical methods (techniques) [11]. The system represents a triangle of food compositions corresponding to 100% content of proteins, carbohydrates, and fats with the vertices of nine sector triangles denoting the content proportions of these food bases. Since the early 2000s, this strategy has been used to develop any RMs of nutritional value, such as products of plant and animal origin, baby and dietary food, dietary supplements, etc. The review of [12] details the evolution of RMs and analytical methods for the determination of organic nutrient content in foods and supplements. Unfortunately, a flat model of a food organic matrix appears to be inapplicable to reflect such a RM feature as the number of certified,

---

[5] GSO 8921–2007 Certified reference material for composition of Canadian pond weed (EK-1). Available at: https://fgis.gost.ru/fundmetrology/registry/19/items/391264.

COOMET CRM 0065–2009-RU CRM for composition of Canadian pond weed (EK-1). Available at: http://www.coomet.org/en/ doc/i2_2008.pdf.

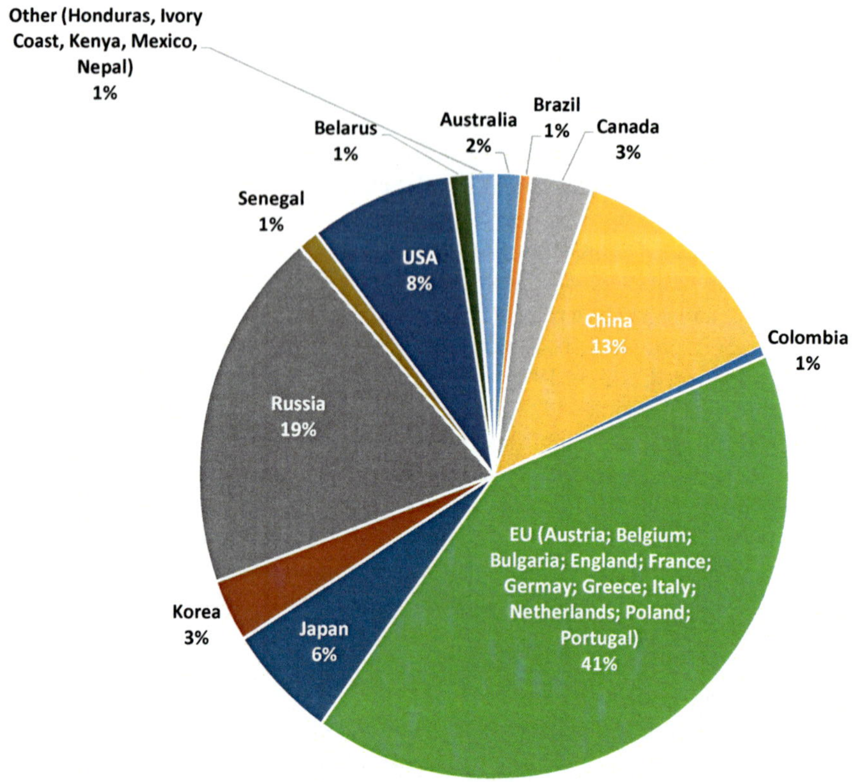

Fig. 3 The number of plant reference materials developed by manufacturers from different countries according to open sources as of 03/01/2020 [9]

recommended, and informational contents for 10–50 chemical elements (so-called "minerals").

B. Markert [13] proposed an original approach to characterizing plants based on their elemental composition, i.e., creation of a "Reference Plant" fingerprint. This direction was further developed by his followers, with chemometrics applied to describe the relationships between the elemental and component compositions of biological plant species [14–16]. Despite the extensive number of publications devoted to the composition of chemical elements and organoelement compounds in different plant species and their organs, chemometric processing methods, providing a significant progress in the systematization of such information, have not yet been found.

Therefore, at present, it appears to be theoretically impossible to indicate the types of plants (or their organs) that should be recommended for preparing plant-matrix RMs, as well as the list of chemical elements that must be certified therein. Nevertheless, the study of the environmental transformation is still relevant due to the observed metallization of the biosphere [17, 18]. Aquatic and terrestrial plants indicate polluting elements and the degree of pollution in their distribution area during environmental monitoring and bioremidation of soils in contaminated areas [19–21].

Geochemical and environmental projects and programs are focused on mapping the distribution of chemical elements in space and time. For such projects, from 30 to 90 biophilic and toxic elements are determined with a wide concentration range in a large number of various species plants. In order to characterize the ecosystem sufficiently completely and reliably by a comparative study, it is necessary to collect many species of different plants, characteristic of the studied territories, following the same protocol for sampling and analytical study of samples and using the same (similar or close in analytical capabilities) equipment. The quality of the results of chemical analysis can be evaluated by encrypted plant-matrix RMs included in the batch of analyzed samples. If the listed conditions are not met, no guarantee is provided to the reliability of analytical data, establishment of reliable CE background concentrations in various plants, as well as correct assessment of the pollution level and risks for the population [22].

Despite the widespread use of medicinal plants in medicine and cosmetology, the mechanisms of their phytotherapeutic action on humans and animals have not been sufficiently studied [23, 24]. Medical and biological projects are devoted to the establishment of medical and hygienic standards of both the maximum permissible concentration (MPC) of toxic elements in water, food, medicinal, cosmetic products, and hygiene products and their permissible safe amount when taken daily with medicinal products. Although plant medicinal and cosmetic preparations have long been used in medical practice, the amount of data on the content of chemical elements in various plants and their organs is rather modest. In order to ensure the safety of their use, it is also necessary to constantly assess the quality of medicinal plant raw materials and medicinal products based on them (decoctions, tinctures, powders and tablets, dietary supplements).

The conventional approach to the study of the chemical composition of plants is associated with agrochemical problems [25, 26], such as increasing the productivity of agricultural crops, the influence of essential and toxic elements on the growth and development of plants, obtaining products with a certain content of organic and mineral substances. Therefore, the determination of the plant chemical composition is carried out in order to assess the safety of food, wild and cultivated medicinal plant raw materials and products from them, evaluate the quality of crop products (premixes of proteins, fats, carbohydrates, ash (mineral) elements, and water), achieve a balance of proteins, carbohydrates, fats, vitamins, biologically active substances, and microelements in the diet of humans, domestic animals, and poultry, monitor the correct use of fertilizers and pesticides, as well as to assess the availability of microelements for agricultural lands. Environmental pollution negatively alters both the productivity of wild and agricultural plants and the nutritional value of foods in the food chain.

A characteristic feature of the conventional approach to the study of the plant chemical composition using different analytical methods consists in discussing and interpreting the results of determining only a small number of biophilic and/or toxic chemical elements. The reliability of analytical data is confirmed by CRMs and RMs, the standard addition method or by comparing the results obtained by several methods of analysis.

In order to determine the CE content, single- and multielement analytical techniques are used, taking into account the peculiarities of plant objects during their selection and sample preparation, such as the concentration range of the determined element in the organic matrix under study. The homogeneity and stability of plant RMs can be characterized and evaluated by direct methods (XRF, INAA, AES with an arc and/or laser excitation, etc.) [27–29] and methods where the dissolution of plant samples is necessary (gravimetry, titrimetry, potentiometry, fluorimetry, spectrophotometry, AAS with a flame or electrothermal atomizer, flame and ICP-AES, ICP-MS, etc.) [21, 24].

Difficulties of direct multielement analytical methods are associated with the fact that calibration requires a large number of matrix RMs with a wide range of analytes certified. Such RM sets can hardly be found. The preparation of plant-matrix sample solutions also has a number of difficulties. These include different solubility of organic and organoelement compounds in mineral and organic acids, the necessity of separating macro- and microelements and group or individual concentration of analytes [30]. In this regard, on the one hand, the list of simultaneously determined elements is reduced, which simplifies the analytical task and does not require the use of plant-matrix RMs of various compositions, since it is often sufficient to use standard solutions for calibration. On the other hand, the analysis time is increased due to the enlarged number of the applied analytical procedures and measurement methods (techniques) for characterizing the product.

Individual plant parts are analyzed when solving problems in agrochemistry and food industry, as well as in the production of pharmaceutical and cosmetic preparations. Quality and safety are assessed for root crops (potatoes, carrots, chicory, ginseng rhizome, etc.), leaves (cabbage, lettuce, spinach, tea, eucalyptus, bay leaves, etc.), stems, stems with leaves, bark (flax, corn, rapeseed, cinnamon, etc.), flowers (roses, tulips, etc.), seeds (crops, sunflowers, legumes, coffee, etc.), and fruits (apples, peaches, dates, strawberries, etc.). Forage crops for animals and poultry, as well as medicinal and culinary raw materials, are often used without separation into roots, stems, leaves, flowers, and fruits. For chemical analysis, plant matter is taken in a living and dried solid state of various moisture content (roots, fruits, berries, grains, leaves, cake or dry powders, tablets, granules), as well as in the form of a liquid (juice, wine, vegetable and essential oils, aqueous and alcohol extracts). Plant samples are submitted for analysis in their natural (living) form, after air drying, sublimation, or lyophilization. Dehydrated plants are usually ground to a powder with a particle size of 0.15 mm or lower. This is why the material for preparing an RM is selected and prepared in the same way.

Many laboratories can perform labor-intensive procedures for the preparation of plant-matrix RMs independently, as well as to determine the content of a large number of elements and estimate their uncertainty budget. However, for most laboratories, using recognized commercial CRMs and RMs appears to be more economically profitable for ensuring consistency of measurements in accordance with ISO 17025–2019

[5–7, 12]. At present, manufacturers have accumulated a great deal of experience and developed guidelines for each stage of the plant-matrix RM development such as: selection, preparation, and homogenization of material, distribution study of elements and compounds; estimation of the smallest representative mass; study of stability, storage conditions, and establishment of the shelf life of the substance; selection of methods of analysis, processing, and presentation of analytical results; algorithms for evaluating certified characteristics, metrological traceability, and consistency of new and previously developed RMs. These recommendations are included in contemporary regulatory documents [31–33] and used in the production of plant-based RMs.

Nevertheless, we have to admit that the insufficient variety of combinations of properties and analyte content, composition and properties of the matrix substance, which results in the discrepancy with the analyzed samples, hampers the existing CRMs and RMs to be fully used by analysts. Under the conditions of the growing demand for traceable and reliable results, which can affect quality of life and cross-border trade, the certification of organic and inorganic pollutants in plants is still relevant. Safe levels of toxic elements are legislated by international and national regulations. The World Health Organization [34] established MPCs for three toxic elements of Cd, As, and Pb in medicinal plant materials equal to 0.3, 1.0, and 10 mg/kg, respectively. The safety requirements for medicinal plant raw materials and drugs based on them, adopted in the Russian Federation [35–37], indicate that the MPC of four chemical elements, the so-called "heavy metals" and arsenic, should not exceed (mg/kg): Pb - 6.0, Cd - 1.0, Hg - 0.1, and As - 0.5. Since 2018, the United States Pharmacopoeial Convention (USP) and the European Union (since May 2019) have introduced new requirements for 24 impurity elements in medicinal plant raw materials and recommended analytical methods for their determination in raw materials, pharmaceutical products, and food additives: Cd, Pb, As, Hg, Co, V, Ni, Tl, Au, Pd, Ir, Os, Rh, Ru, Se, Ag, Pt, Li, Sb, Ba, Mo, Cu, Sn and Cr [38–40]. In this regard, new and additional certification should be developed for existing plant-matrix RMs, for the validation and assessment of the uncertainty of existing and new methods of analysis. The successful application of RMs ensures quality control of results, certification studies, and proficiency tests.

Based on the generalization of international and Russian regulatory documents, publications, as well as our own experimental studies on the development and use of RMs, a list of the most important requirements for the procedures of preparation, characterization, and certification of new plant-matrix RMs was compiled:

(1) the substance must be prepared in such a way as to ensure its long-term storage without degradation (decomposition) of organic compounds;
(2) the prepared material must be a homogeneous mixture;
(3) certification should be based on the results of at least two analytical methods, taking into account their traceability to SI units and consistency of national collections of different countries;
(4) the amount of RM should be sufficient and available for long-term use in various analytical methods;

(5) the list of certified elements should be formed taking into account the degree of their toxicity in the case of using RM for assessing the safety of agricultural products, medicinal plants, pharmaceuticals, and cosmetics.

The latter requirement is reflected in the Pharmacopoeias of different countries and is constantly evolving in connection with the development of instrumental methods of chemical analysis.

Conclusion

Over the past few decades, since the quality has become a critical issue in various areas of life, the importance of RMs in chemical metrology has increased dramatically. The worldwide demand for traceable and reliable results is continuously growing and the problem of obtaining traceable and reliable analytical results is still relevant.

In accordance with the requirements of ISO 17025–2019, plant RMs represent a means for ensuring the uniformity of chemical measurements in geochemistry, ecology, agriculture, and pharmacology.

Currently, a large list of generally recognized commercial matrix CRMs and RMs of plants and products based thereon, produced by different manufacturers, is available. Therefore, CRMs and RMs with metrological characteristics prescribed in the certificates are used by analytical laboratories. Nevertheless, the existing variety of plant-matrix RMs appears to be insufficient in terms of the number of types and certified chemical elements to control the quality of analytical procedures and the results of chemical analytical methods used in industrial spheres and scientific research.

About 30–40% of publications on the elemental compositions of plants contain no information on assessing the reliability of analytical data. This is most typical of Russian articles on the elemental and component composition of particular plant species, despite the fact that the majority of contemporary commercial and research analytical laboratories are accredited with the quality assurance and control procedures, which are carried out using RMs. In addition, it must be acknowledged that the diversity of "CE/content level/composition and properties of the plant matrix" combinations in existing CRMs and RMs hampers selection of RMs fully corresponding to the composition and properties of the analyzed samples.

We have generalized the existing requirements and criteria for RMs of plant origin. On the basis of literature data and own experimental studies, we compiled a list of the most important requirements for plant-matrix RMs aimed at obtaining reliable results by various methods of chemical analysis for ensuring safety of food, drugs and, in general, the environment.

Acknowledgements The study was carried out within the framework of the state assignment for Projects IX.127.1.4. No. 0350-2019-0005 "Ecological and geochemical transformations of ecosystems in Eastern Siberia under the influence of natural and technogenic factors" and No. 0284-2021-0005 "Development of methods for studying the chemical composition and structural state of natural and technogenic environments in the earth sciences."

Author Contributions All authors have made an equivalent contribution to the preparation of the publication.

Conflict of Interest The article was prepared on the basis of a report presented at the IV International Scientific Conference "Reference Materials in Measurement and Technology" (St. Petersburg, December 1–3, 2020). The article was admitted for publication after the abstract was revised, the article was formalized and the review procedure was carried out.

The version in the Russian language is published in the journal "Measurement Standards. Reference Materials" 2021;17(2):33–47. https://doi.org/10.20915/2687-0886-2021-17-2-33-47.

References

1. 2010.1. IUCN red list of threatened species: summary statistics. In: International union for conservation of nature and natural resources. https://www.iucnredlist.org
2. Katz SA (2002) Bowen's Kale: A brief review dedicated to the late Professor Humphry John Moule Bowen, 1929–2001. J Radioanal Nucl Chem 251(1):3–5. https://doi.org/10.1023/A:1015021823497
3. RF Federal law (2008) On ensuring the uniformity of measurements. No FZ-102 of 26/06/2008. (In Russ.). Available at: http://fundmetrology.ru/depository/01_npa/102-fz_2015.pdf
4. GOST ISO/IEC 17025–2019 (2020) General requirements for the competence of testing and calibration laboratories. Standartinform, Moscow, p 13. (In Russ.)
5. Linsinger TPJ, Emons H (2009) The role of reference materials in chemical metrology. CHIMIA Int J Chem 63(10):629–631. https://doi.org/10.2533/chimia.2009.629
6. Vasil'eva IE, Shabanova EV (2017) Certified reference materials of geological and environmental objects: problems and solutions. J Analyt Chem 72(2):99–118. https://doi.org/10.1134/S1061934817020149
7. Olivares IRB, Souza GB, Nogueira ARA, Toledo GTK, Marcki DC (2018) Trends in developments of certified reference materials for chemical analysis - focus on food, water, soil, and sediment matrices. TrAC, Trends Anal Chem 100:53–64. https://doi.org/10.1016/j.trac.2017.12.013
8. Hulme N, Hammond J (2020) Is your spectrophotometer still "pharma compliant"? A review of the new European pharmacopoeia 10th edition. SPECTROSCOPYEUROPE 32(1):14–20. https://www.spectroscopyeurope.com/article/your-spectrophotometer-still-"pharma-compliant"-review-new-european-pharmacopoeia-10th
9. Vasil'eva IE, Shabanova EV (2021) Plant-matrix certified reference materials as a tool for ensuring the uniformity of chemical measurements. J Analy Chem 76(2):137–155. https://doi.org/10.1134/S1061934821020143
10. Catalogue of certified reference materials of natural and man-made media compositions from A.P. Vinogradov Institute of Geochemisty SB RAS. Available at: http://www.igc.irk.ru/images/Innovation/Standarts-obr/CATALOGUE_OF_CRMs_IGC_SB_RAS_-2017.pdf
11. Wolf WR, Andrews KW (1995) A system for defining reference materials applicable to all food matrices Fresenius. J Anal Chem 352(1–2):73–76. https://doi.org/10.1007/BF00322300
12. Wise SA, Phillips MM (2019) Evolution of reference materials for the determination of organic nutrients in food and dietary supplements – a critical review. Anal Bioanal Chem 411(1):97–127. https://doi.org/10.1007/s00216-018-1473-0

13. Markert B (1992) Establishing of "Reference Plant" for inorganic characterization of different plant species by chemical fingerprinting. Water, Air, Soil Pollut 64(3–4):533–538. https://doi.org/10.1007/BF00483363
14. Owen JD, Kirton SB, Evans SJ, Stair JL (2016) Elemental fingerprinting of Hypericum perforatum (St John's Wort) herb and preparations using ICP-OES and chemometrics. J Pharm Biomed Anal 125(5):15–21. https://doi.org/10.1016/j.jpba.2016.02.054
15. Arsenijević J, Marković J, Šoštarić I, Ražić S (2013) A chemometrics as a powerful tool in the elucidation of the role of metals in the biosynthesis of volatile organic compounds in Hungarian thyme samples. Plant Physiol Biochem 71(10):298–306. https://doi.org/10.1016/j.plaphy.2013.08.002
16. Habte G, Hwang IM., Kim JS, Hong JH, Hong YS, Choi JY et all (2016) Elemental profiling and geographical differentiation of Ethiopian coffee samples through inductively coupled plasma-optical emission spectroscopy (ICP-OES), ICP mass spectrometry (ICP-MS) and direct mercury analyzer (DMA). Food Chem 212(Dec 1):512–520. https://doi.org/10.1016/j.foodchem.2016.05.178
17. Jurkin D, Zgorelec Z, Rinkovec J (2019) Concentrations of Pt, Pd and Rh in soil and vegetation: A review. J Centr Eur Agricul 20(2):686–699. https://doi.org//https://doi.org/10.5513/JCEA01/20.2.2199
18. Nagajyoti PC, Lee KD, Sreekanth TVM (2010) Heavy metals, occurrence and toxicity for plants: a review. Environ Chem Lett 8(3):199–216. https://doi.org/10.1007/s10311-010-0297-8
19. Reimann C, Koller F, Frengstad B, Kashulina G, Niskavaara H, Englmaier P (2001) Comparison of the element composition in several plant species and their substrate from a 1500000-km2 area in Northern Europe. Sci Total Environ 278(1–3):87–112. https://doi.org/10.1016/S0048-9697(00)00890-1
20. Nečemer M, Kump P, Ščančar J, Jaćimović R, Simčič J, Pelicon P, Budnar M, Jeran Z, Pongrac P, Regvar M, Vogel-Mikuš K (2008) Application of X-ray fluorescence analytical techniques in phytoremediation and plant biology studies. Spectrochim Acta, Part B 63(11):1240–1247. https://doi.org/10.1016/j.sab.2008.07.006
21. Kroukamp EM, Wondimu T, Forbes PBC (2016) Metals and metalloids speciation in plants: overview, instrumentation, approaches and commonly assessed elements. Trends Anal Chem 77:87–99. https://doi.org/10.1016/j.trac.2015.10.007
22. Eggen OA, Reimann C, Flem B (2019) Reliability of geochemical analyses: deja vu all over again. Sci Total Environ 670:138–148. https://doi.org/10.1016/j.scitotenv.2019.03.185
23. Khan A, Khan S, Khan MA, Qamar Z, Waqas M (2015) The uptake and bioaccumulation of heavy metals by food plants, their effects on plants nutrients, and associated health risk: a review. Environ Sci Pollut Res 22(18):13772–13799. https://doi.org/10.1007/s11356-015-4881-0
24. Pohl P, Bielawska-Pohl A, Dzimitrowicz A, Greda K, Jamroz P, Al L et al (2018) Understanding element composition of medicinal plants used in herbalism – a case study by analytical atomic spectrometry. J Pharm Biomed Anal 159:262–271. https://doi.org/10.1016/j.jpba.2018.06.017
25. Mengel K, Kirkby EA, Kosegarten H, Appel T (2001) Principles of Plant Nutrition. 5th ed. Springer Netherlands, Dordrecht, p 849. https://doi.org/10.1007/978-94-010-1009-2
26. Temminghoff EEJM, Houba VJG (2004) Plant Analysis Procedures. 2nd Ed. Springer Netherlands, Dordrecht, p 179. https://doi.org/10.1007/978-1-4020-2976-9
27. Greenberg RR, Bode P, Fernandes EAN (2011) Neutron activation analysis: A primary method of measurement. Spectrochim Acta, Part B 66(3–4):193–241. https://doi.org/10.1016/j.sab.2010.12.011
28. Vanhoof C, Bacon JR, Ellis AT, Fittschen UEA, Vincze L |(2019) 2019 atomic spectrometry update – a review of advances in X-ray fluorescence spectrometry and its special applications. J Analy Atomic Spect 34(9):1750–1767. https://doi.org/10.1039/C9JA90042J
29. Senesi GS, Cabral J, Menegatti CR, Marangoni B, Nicolodelli G (2019) Recent advances and future trends in LIBS applications to agricultural materials and their food derivatives: An overview of developments in the last decade (2010–2019). Part II. Crop plants and their food derivatives. Trends Anal Chem 118:453–469. https://doi.org/10.1016/j.trac.2019.05.052

30. Tsizin GI, Statkus MA, Zolotov YuA (2015) Adsorption and extraction preconcentration of trace components in flow analytical systems. J Anal Chem 70(11):1289–1306. https://doi.org/10.1134/S1061934815110167
31. ISO Guide 33:2015 (2015) Reference materials. Good practice in using reference materials. BSI, Geneva. Available at: https://www.iso.org/standard/46212.html
32. ISO/IEC Guide 98–3:2008 (2008) Uncertainty of measurement — Part 3: Guide to the expression of uncertainty in measurement (GUM:1995). ISO, Geneva. Available at: https://www.iso.org/standard/50461.html
33. ISO Guide 35–2015 (2016) Reference materials – General and statistical principles of certification. Moscow; Standartinform (In Russ.)
34. World Health Organization (2005) National policy on traditional medicine and regulation of herbal medicines: Report of a WHO global survey. Geneva
35. State Pharmacopoeia of the Russian Federation, XIII edn. Moscow, 2016. Available at: https://www.rosminzdrav.ru/poleznye-resursy/gosudarstvennaya-farmakopeya-rossiyskoy-federatsii-xiii-izdaniya
36. State Pharmacopoeia of the Russian Federation, XIV edn. Moscow, 2018. Available at: http://www.femb.ru/femb/pharmacopea.php
37. Resolution of the chief state sanitary doctor of the Russian Federation "On the introduction of sanitary rules" dated November 14, 2001 No 36 (as amended on July 6, 2011). SanPiN 2.3.2.1078–01. Hygienic requirements for food safety and nutritional value. Available at: https://base.garant.ru/4178234/
38. United States Pharmacopeia. General Chapter <232> Elemental Impurities – Limits: First Supplement of USP 40-NF35, Official December 1, 2017. Available at: https://www.usp.org/sites/default/files/usp/document/our-work/chemical-medicines/key-issues/232-40-35-1s.pdf
39. United States Pharmacopeia. General Chapter <233> Elemental Impurities – Procedures / Chemical Tests: Second Supplement to USP 38–NF 33. Available at: https://www.usp.org/sites/default/files/usp/document/our-work/chemical-medicines/key-issues/c233.pdf
40. ICH International Council for Harmonisation of Technical Requirement for Pharmaceuticals for Human Use, Harmonised Guideline, Guideline for Elemental Impurities Q3D (R1) (Final version Adopted on 22 March 2019). Available at https://www.ema.europa.eu/e/documents/scintific-guideline/international-conference-harmonisation-technical-requirements-registration-pharmaceuticals-human-use_en-32.pdf^m

Irina E. Vasil'eva Dr. Sci. (Eng.), Chief Researcher of the group of atomic emission analysis methods and reference materials, A. P. Vinogradov Institute of Geochemistry, Siberian Branch of the Russian Academy of Sciences (IGC SB RAS).

Elena V. Shabanova Dr. Sci. (Phys.-Math.), Senior Researcher, the head group of atomic emission analysis methods and reference materials, A.P. Vinogradov Institute of Geochemistry, Siberian Branch of the Russian Academy of Sciences (IGC SB RAS).

Reference Materials for Optical Nanosensor Systems: Reduced Glutathione and Chloramphenicol

Anna A. Yushina and Mikhail K. Alenichev

Abstract The research presents an approach to developing in-house reference materials (IHRMs) for a nanosensor system based on dynamic light scattering and fluorescence for qualitative and quantitative determination of the food contaminant antibiotic chloramphenicol (laevomycetin) and reduced glutathione, a marker for ischemic stroke and several other diseases. Chloramphenicol and reduced glutathione were chosen as candidate materials. The certification procedure based on the calculation and experimental method of preparation was employed to establish the certified value of the IHRM. During the tests, the metrological characteristics of the reference material were determined. The certified value of the mass fraction of the IHRM for reduced glutathione is 98.5%, and the expanded uncertainty of the certified value with the coverage factor $k = 2$ is $\pm$ 0.3%. The certified value of the mass concentration of the IHRM for chloramphenicol is 10.0 g/dm^3, and the expanded uncertainty of the certified value with the coverage factor $k = 2$ is $\pm$ 4.0%. The use of the developed IHRMs demonstrated their applicability for calibration of optical nanosensor systems based on dynamic light scattering and fluorescence. It is assumed that the developed in-house reference materials can be further certified as certified reference materials (CRM) and used for verification, calibration, and graduation of compact detection devices of the "point of care diagnostics" type designed for express tests right on the sampling site.

Keywords Reference material · Optical nanosensor system · Disease marker · Food contaminant · Dynamic light scattering · Fluorescence

A. A. Yushina (✉) · M. K. Alenichev
All-Russian Scientific Research Institute for Optical and Physical Measurements (VNIIOFI), Ozernaya st. 46, Moscow 119361, Russia
e-mail: yushina@vniiofi.ru

M. K. Alenichev
e-mail: alenichev@vniiofi.ru

S. V. Medvedevskikh et al. (eds.), *Reference Materials in Measurement and Technology*,
https://doi.org/10.1007/978-3-031-06285-8_16

Introduction

The development direction of optical nanosensor systems has been burgeoning lately as the unique optical properties of nanoscale objects permit implementing highly sensitive detection of chemical substances, microorganisms, and biological molecules, including chemical contaminants of food, environmental media, and disease markers [1]. In many cases such systems provide a means of performing rapid tests directly on the sampling site (i.e. point of care diagnostics). The basis for optical nanosensors is the measurement of light absorption, fluorescence, or light scattering by nanoparticles interacting with the analyte [2].

Optical nanosensor systems based on the surface plasmon resonance effect [3] and various fluorescence effects such as fluorescence resonance energy transfer (FRET) [4, 5], the effect of excitation energy transfer from fluorophore [6, 7], or the inner filter effect [8, 9] are particularly common. In addition, there is a large amount of nanosensor systems based on dynamic light scattering (DLS) [10–12]. Nanosensor systems intended for quantitative analysis require the construction of calibration curves. For this purpose, reference materials of corresponding analytes are necessary. The requirements specified for reference materials for nanosensor systems include the correspondence of concentration ranges, the preparation simplicity of working solutions from standard reference materials, stability, purity, resistance to external influences, and culture media coordination. Ideally, the matrix of the standard reference material should correspond to the matrix of the sample which is going to be analysed later. Unfortunately, this condition is not always realizable in actual practice.

Optical nanosystems based on DLS and fluorescence were designed in the All-Russian Research Institute for Optical and Physical Measurements (VNIIOFI, Moscow) [1, 13]. The system based on DLS was developed employing the competition assay approach. It is aimed at detecting the antibiotic chloramphenicol (whose trade name is laevomycetin), an antimicrobial medicinal product, which is a dangerous contaminant of food, drinking water, and wastewater [14]. The fluorescent nanosystems were designed to detect and quantify reduced glutathione, a marker for ischemic stroke and several other diseases. These systems are based on the FRET effect and the effect of quantum dots excitation energy transfer. To provide metrological support of the measurements taken with optical nanosensor systems based on DLS and fluorescence, it was necessary to develop reference materials meeting the requirements formulated above and permitting the reproduction of mass concentrations of reduced glutathione and chloramphenicol.

This paper is an investigation into the feasibility of developing in-house reference materials (IHRM) intended for the metrological support of measurement procedures for chloramphenicol and reduced glutathione. The certified values of the mass concentrations of these analytes were determined through the mass measurement results obtained with an analytical balance traceable to the state primary reference material (SPRM) of mass and volume. The measurements were performed with the help of dispensers traceable to the SPRM of fluid volume. The metrological characteristics of the precision balances and dispensers available at the All-Russian Research Institute for Optical and Physical Measurements (VNIIOFI) enable providing expanded uncertainties of the certified values of mass concentration which do not exceed 4.0% for chloramphenicol and 0.3% for reduced glutathione.

Materials and Methods

Characteristics of the Certified Reference Materials

Reduced Glutathione

Glutathione (γ-L-Glutamyl-*L*-cysteinylglycine) is a tripeptide (γ-glutamyl-*L*-cysteinyglycine) consisting of glutamic acid residues, cysteine, and glycine. It is a white crystalline powder or colorless crystals and is highly soluble in water and practically insoluble in methanol and diethyl ether [15].

Commercially available Sigma-Aldrich product CAS No. 70–18-8, cat. no. PHR1359[1] with at least 98.5% purity was chosen as the candidate material for preparing the IHRM for reduced glutathione. It is a secondary reference material developed in accordance with ISO 17034 [16] and ISO/IEC 17,025 [17]. The secondary reference material was prepared gravimetrically using a balance certified and calibrated as specified in the ISO 17025 requirements. All calibrations used NIST traceable weights calibrated in compliance with the NIST standards in a laboratory accredited according to ISO 17025. Metrological traceability to the corresponding primary standard is achieved via direct comparison.

No research of the uniformity estimation of the candidate for the IHRM was performed as its application involves single dissolution in deionized water and utilization of the entire quantity of the IHRM material from the test tube.

Chloramphenicol

Chloramphenicol (Chloramphenicolum) is an antibiotic in the form of a white crystalline powder, thin crystals, or oblong plates. It is highly soluble in ethanol, soluble in ethyl acetate, and poorly soluble in water [18].

---

[1] Pharmaceutical Secondary Standard; Certified Reference Material Glutathione 70–18-8. https://www.sigmaaldrich.com/catalog/product/sial/phr1359?lang=en&region=RU.

Commercially available Sigma-Aldrich product CAS No. 56–75-7, cat. no. C0378[2] with at least 98.0% (99%) purity was chosen as the candidate material for developing the IHRM for chloramphenicol.

Ethanol complying with State Standard (GOST) 5962–2013 was chosen as the solvent [19]. Since chloramphenicol molecules are evenly distributed throughout the entire solution volume due to Brownian diffusion in liquids and the identical charges of molecules, which is caused by the high solubility of chloramphenicol in ethanol, close distances between chloramphenicol molecules are ensured. For this reason, no research of the uniformity estimation of the candidate material for the IHRM was performed. The IHRM, consequently, is a true solution; there are no sources of uncertainly connected with its inhomogeneity.

For these reference materials the category "in-house reference material" was selected.

Determination of the Certified Values of the Reference Materials and Estimation of the Expanded Uncertainty of the Certified Values

Reduced Glutathione

The certified value of the IHRM was determined by the manufacturer (accredited in accordance with ISO/IEC 17,025) of the material chosen for the IHRM. The certification procedure based on the application of the reference material for reduced glutathione of the USA pharmacopoeia (the US glutathione standard) was employed to estimate the certified mass fraction value of the IHRM for reduced glutathione.

High-performance liquid chromatography was used to determine the certified mass fraction value of reduced glutathione.

Arrangements for measurements:

- Column: Ascentis Express C18, 4.6 × 50 mm, 2.7 μm
- Mobile phase: 6.8 g/l of K_2HPO_4 and 2.02 g/l of sodium heptanesulphonate in water (pH 3.3)–methanol (96:4)
- Flow rate: 1 ml/min
- Temperature: 30°C
- Sample volume: 3 μl
- Detection wavelength: 210 nm

In accordance with this certification procedure, the reference standard for glutathione of the United States Pharmacopoeia (USP) USP LOT R106J0 with the

[2] Pharmaceutical Secondary Standard; Certified Reference Material Glutathione 70-18-8. https://www.sigmaaldrich.com/catalog/product/sial/phr1359?lang=en&region=RU (Accessed 15 April 2021).

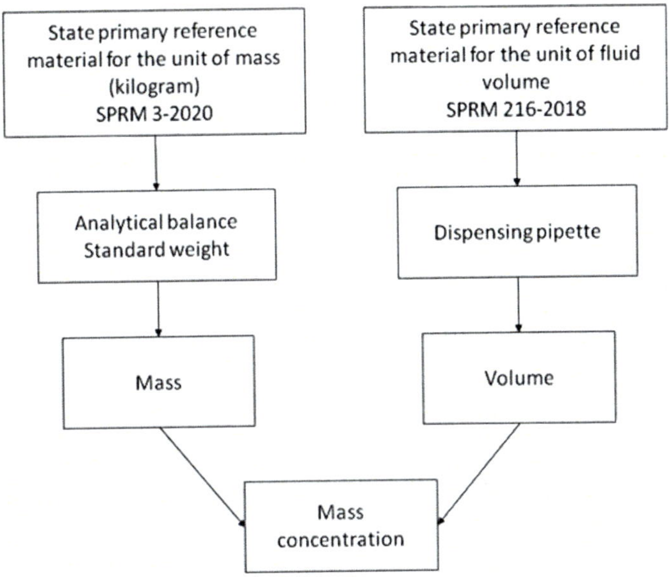

Fig. 1 Diagram of the metrological traceability of chloramphenicol mass concentration measurements

reduced glutathione content of 0.99 mg/mg was used for a preliminary calibration of a high-performance liquid chromatograph. The calibration was followed by a mass fraction measurement of the reduced glutathione in the material of the IHRM under development.

The metrological traceability of the IHRM certified value is estimated in relation to the reduced glutathione reference material of the United States Pharmacopoeia.

Chloramphenicol

The certified value of the candidate material for developing the IHRM for chloramphenicol was established by calculation according to the preparation procedure. The traceability of the mass concentration values was ensured using instruments for mass measurement (an analytical balance) traceable to the state primary reference material of the unit of mass SPRM 3–2020[3] and instruments for measuring the volume of fluid (a dispenser) traceable to the state primary reference material of the unit of fluid volume SPRM 216–2018.[4] The diagram in Fig. 1 shows the metrological traceability of the measurement results of chloramphenicol mass concentration.

[3] SPRM 3–2020 State primary reference material of the unit of mass (kilogram). https://fgis.gost.ru/fundmetrology/registry/12/items/1385582.

[4] SPRM 216–2018 State primary reference material of the unit of fluid volume in the range of 1.0·10-9 m^3 to 1.0 $м^3$. https://fgis.gost.ru/fundmetrology/registry/12/items/397905.

To determine the certified value of the IHRM for chloramphenicol mass concentration, the certification procedure based on the calculation and experimental method of preparation [20] was employed. The following measuring instruments were used to establish the certified values of chloramphenicol mass concentration:

- General-purpose weighing scale of class 1 accuracy to State Standard GOST OIML R 76-1-2011 [21] with the maximum permissible absolute measurement error of a single weighing not more than ±0.0001 g.
- One-milligram standard weight[5] as per State Standard GOST OIML R 111-1-2009 [22].
- Adjustable volume dispensing pipette covering ranges of 1000 to 10,000 mcl.[6] The maximum permissible systematic component of the basic relative error is ±1.0%. The maximum permissible mean-square deviation of the random component of relative error is 1.0%.
- Microclimate tester Meteoskop-M.[7]

A 0.05 chloramphenicol weighed sample was placed into a pre-weighed glass weighing bottle with a cap with a capacity of at least 10 cm^3. Then 5 cm^3 of ethanol was added using an adjustable dispensing pipette. The weighing bottle was capped, and its content was thoroughly mixed until the chloramphenicol completely dissolved in the ethanol. After that, the resulting solution was put into a container for storing the IHRM.

To measure the mass of the chloramphenicol weighed sample and the non-dissolved remainder of the chloramphenicol in the glass weighing bottle with a cap, the weighing procedure was performed as follows:

The weighing bottle with a cap was placed onto the weighing pan and weighed five times. A one-milligram standard weight was added to the same weighing pan (the actual mass was 1.001 mg) and weighed together with the capped weighing bottle five times.

The sensitivity of one reading of the scales (δm) was calculated according to Eq. (1):

$$\delta m = \frac{1mg}{(n_{b+w} - n_b)}, \tag{1}$$

where n_{b+w} is the mean number of discrete readings while weighing a capped weighing bottle with the standard weight; n_b is the mean number of discrete readings while weighing a clean weighing bottle.

---

[5] Class E1, E2, F1, F2, and M1 weights (number in the state register 36068-07). https://fgis.gost.ru/fundmetrology/registry/4/items/345346.

[6] Single- and multi-channel adjustable pipettes (number in the state register 37432-13). https://fgis.gost.ru/fundmetrology/registry/4/items/346979.

[7] Microclimate testers (number in the state register 32,014–11). https://fgis.gost.ru/fundmetrology/registry/4/items/340205.

The standard weight was removed from the weighing pan. The chloramphenicol weighed sample was placed into the weighing bottle with a cap and weighed five times.

The mass of the chloramphenicol weighed sample (m_s) was calculated according to Eq. (2):

$$m_s = (n_{b+s} - n_b) \cdot 1mg, \tag{2}$$

where n_{b+s} is the scale reading while weighing the capped weighing bottle and the chloramphenicol weighed sample, mg; n_b is the scale reading while weighing the weighing bottle with a cap and without chloramphenicol, mg.

The mass concentration of chloramphenicol in the C solution (mg/dm^3) was calculated using Eq. (3):

$$C = \frac{1000 \cdot (m_s - m_r)}{V}, \tag{3}$$

where m_s is the mass of the chloramphenicol weighed sample taken to prepare the IHRM, g; m_r is the mass of the chloramphenicol non-dissolved residue, g; V is the volume of the ethanol taken to prepare the IHRM, cm^3.

The mass of the non-dissolved IHRM residue in the weighing bottle was determined using Eq. (2) and following the procedure described above.

The quantitative evaluation of the standard uncertainty of the IHRM certified value relates to the determination of the chloramphenicol weighed sample mass and the volume of ethanol used to prepare the IHRM.

According to the type specification, the systematic weighing error of the laboratory analytical balance (Ohaus Explorer Pro, model EP114C, accuracy class I, 0.1 mg resolution[8]) is 0.75 mg. The additive component of the systematic error was eliminated with the error compensation by sign method. This method implies taking measurements in such a way that the measurement error would enter the data with one sign first and with the opposite sign next time. In accordance with this premise, to determine the mass of the chloramphenicol weighed sample and the non-dissolved residue, two weighings of each were performed: clean laboratory ware was weighed during the first weighing and laboratory ware with the chloramphenicol weighed sample or the non-dissolved residue during the second one. The chloramphenicol weighed sample mass was determined as the difference between the two above-mentioned weighings. As a result, the additive component of the systematic error was eliminated. The non-eliminated (multiplicative) component of the systematic error was taken equal to the resolution of the balance, i.e. 0.1 mg. Thus, the expanded uncertainty of the B type connected with the technical characteristics of the balance was taken equal to the non-eliminated component of the systematic error. Due to

[8] Electronic balance (number in the state register 16,313–08). https://fgis.gost.ru/fundmetrology/registry/4/items/315230.

this, the standard uncertainty of the B type was estimated according to Eq. (4):

$$u_B = \frac{d}{1,96}, \tag{4}$$

where d is the resolution of the used balance, i.e. 0.1 mg.

The standard uncertainty of the A type associated with weighing the chloramphenicol weighed sample was evaluated according to Eq. (5):

$$u_A = \sqrt{\frac{\sum_{i=1}^{n} (\mathrm{m_m} - \mathrm{m}_{am})^2}{n(n-1)}}, \tag{5}$$

where m_m is the results of mass measurements of the chloramphenicol weighed sample, mg; m_{am} is the arithmetic mean value of the chloramphenicol weighed sample mass obtained from the n measurements, mg; n is the number of mass measurements of the weighed chloramphenicol sample.

The standard uncertainty (u_V) associated with the ethanol volume was evaluated using the data from the calibration certificate of the utilized dispensing pipette and calculated according to Eq. (6):

$$u_V = \sqrt{u_A^2 + u_B^2}, \tag{6}$$

where u_A is the type A standard uncertainty taken equal to the maximum permissible mean-square deviation of the random component of relative error, cm^3; u_B is the B type standard uncertainty taken equal to the maximum permissible systematic component of the basic relative error, cm^3.

According to the type specification of the adjustable volume dispensing pipette covering ranges of 1000 to 10,000 mcl[6], the maximum permissible mean-square deviation of the random component of relative error is 1.0%; the maximum permissible systematic component of the basic relative error is ± 1.0%. The values of the A and B types of the standard uncertainty are taken equal to the maximum permissible mean-square deviation of the random component of relative error and the maximum permissible systematic component of the basic relative error in measuring the ethanol volume, respectively.

The combined standard uncertainty of the IHRM certified value in function of the technique used to estimate the certified value of the IHRM (u_t) was evaluated using the equation:

$$u_t = 100 \cdot \sqrt{\left(\frac{1,41 \cdot u_B}{m_s - m_r}\right)^2 + \left(\frac{u_V}{V}\right)^2}, \tag{7}$$

where u_t is the combined standard uncertainty of chloramphenicol mass concentration in the IHRM, %; u_B is the standard uncertainty of mass measurement during weighing, mg; m_s is the mass of the weighed chloramphenicol sample taken for the preparation of the IHRM, mg; m_r is the mass of the non-dissolved chloramphenicol residue, mg; u_V is the standard uncertainty of volume measurement dependent on the dispensing pipette, cm^3; V is the added volume of the ethanol taken to prepare the suspension, cm^3.

The combined standard uncertainty of the certified value (u) was taken equal to the standard uncertainty in function of the technique employed to determine the IHRM certified value:

$$u = u_t, \tag{8}$$

where u_t is the standard uncertainty in function of the technique used to estimate the certified value of the IHRM, %. The expanded uncertainty (with the coverage factor k = 2) of the IHRM certified value was evaluated in accordance with Eq. (9):

$$U_E = 1,96 \cdot u. \tag{9}$$

The calculation results of the combined standard and expanded uncertainties are shown in Table 1.

Stability testing of the materials for both IHRMs was not performed as the established validity period for the IHRMs being developed is much shorter than the established expiry validity period for the source materials. The shelf life of the IHRM for reduced glutathione is six months, while the expiry date of the IHRM material is three years. The shelf life of the IHRM for chloramphenicol is one month, while the expiry date of the IHRM material (chloramphenicol) is four years.

Table 1 Uncertainty characterization of the IHRM for chloramphenicol

Characterization of the IHRM uncertainty	Calculation data, %
Standard uncertainty dependent on the technique of determining the IHRM certified value	2.0
Combined standard uncertainty	2.0
Expanded uncertainty of the certified value with the coverage factor k = 2	4.0

Table 2 Metrological characterization of the developed IHRMs

In-house reference material	Name of the characteristic being certified	Certified value	Expanded uncertainty of the certified value with k = 2, %
Reduced glutathione	Mass fraction, %	98.5	± 0.3
Chloramphenicol	Mass concentration, g/dm^3	10.0	± 4.0

Results and Discussion

Reduced Glutathione

An IHRM for glutathione mass fraction has been developed to reproduce the mass concentration of reduced glutathione. The metrological characteristics of the IHRM for reduced glutathione are the mass fraction of reduced glutathione in the composition of the IHRM and the value of the expanded uncertainty of the certified value with the coverage factor k = 2 (Table 2). The IHRM is a powder with an analyte mass fraction equal to 98.5%; it is intended for the preparation of preset concentration working solutions immediately before use. Deionized water is utilized as a solvent. The IHRM is intended for verification, calibration, and graduation of measuring instruments as well as controlling metrological characteristics during their testing. The IHRM can also be used to control the accuracy of measurement results and to certify measurement procedures used at the enterprise. Additionally, the IHRM is intended for detecting reduced glutathione mass concentration in water and other liquid and biological media with an optical nanosensor system based on fluorescence.

In the first place, the developed IHRM is designed for plotting calibration curves while detecting reduced glutathione in unknown samples as there is a direct correlation between reduced glutathione concentration and the fluorescence intensity of quantum dots in the nanosensor systems developed in VNIIOFI. For instance, the calibration graph given in Fig. 2 was constructed for measuring reduced glutathione concentration in human blood plasma samples with a system based on the effect of quantum dots excitation energy transfer. The designed system provides a means for determining reduced glutathione in the range of 0 to 100 $\mu mol/dm^3$, which corresponds to the physiological concentration of glutathione in blood plasma.

Chloramphenicol

An IHRM was developed to reproduce chloramphenicol mass concentration. The reference material has the following metrological characteristics: the mass concentration of chloramphenicol in ethanol and the value of the expanded uncertainty of the certified value with the coverage factor k = 2 (Table 2). The IHRM is an ethanol

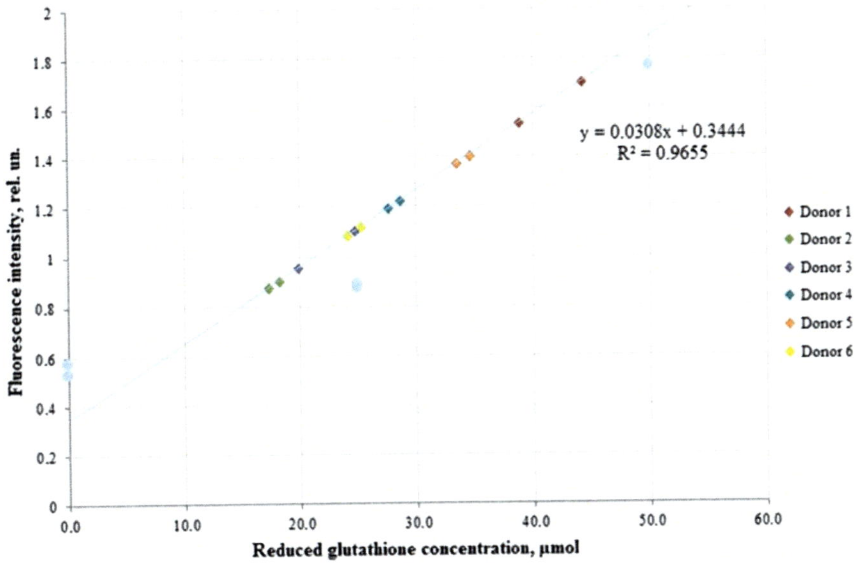

Fig. 2 Calibration graph and the values of reduced glutathione concentration in plasma determined with a nanosensor system based on the effect of quantum dots excitation energy transfer

solution of chloramphenicol with a mass concentration of 10.0 g/dm^3. The IHRM is assumed to be diluted to prepare working solutions of preset concentrations immediately before use. The IHRM purposes are as follows: verification, calibration, and graduation of measuring instruments as well as controlling metrological characteristics during their testing. The IHRM can also be utilized to control the accuracy of measurement results and to certify measurement procedures followed at the enterprise. Additionally, the IHRM is intended for determining chloramphenicol mass concentration in water and liquid and biological media including foodstuffs, food raw material, environmental media.

Figure 3 shows an example of a calibration graph constructed for measuring chloramphenicol mass concentration using the DLS method and a nanosensor system based on competition assay developed by VNIIOFI.

Conclusion

The present paper focused on the feasibility of developing in-house reference materials intended for metrological support of the procedures for measuring chloramphenicol and reduced glutathione with optical nanosensor systems, for detecting the antibiotic chloramphenicol (whose trade name is laevomycetin), for the detection and quantification of reduced glutathione, a marker for ischemic stroke and several other diseases. The certified value of the IHRM for chloramphenicol is traceable to

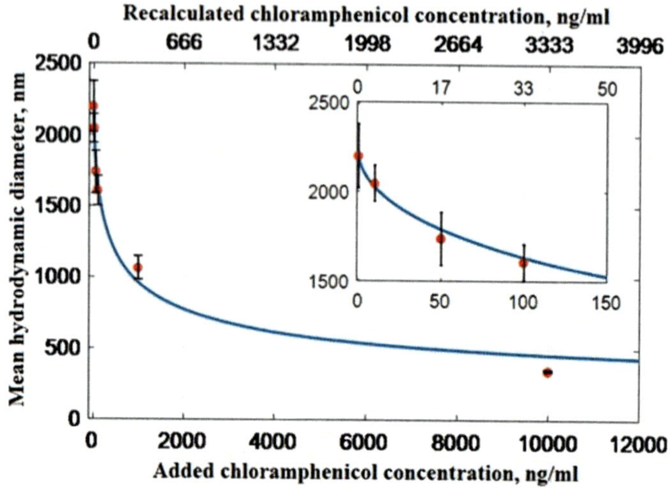

Fig. 3 Size dependence of functionalized gold nanoparticles on chloramphenicol concentration in aqueous solution

state primary reference materials SPRM 3–2020 and SPRM 216–2018. The certified value of the IHRM for reduced glutathione is traceable to the certified reference material for reduced glutathione of the USA pharmacopoeia.

The developed IHRMs for chloramphenicol and reduced glutathione can be further certified as certified reference materials and used for verification, calibration, and graduation of compact detection devices of the "point of care diagnostics" type designed for express tests right on the sampling site.

Acknowledgements The authors express their gratitude to Alexander Davidovich Levin, Dr. Sci. (Engineering), a Senior Researcher of the Laboratory of Optical Spectral Devices at VNIIOFI, for the performed calculations and assistance with the preparation of the manuscript for publication. All measurements were made using the equipment of VNIIOFI.

Author Contributions A.A. Yushina: literature review; preparation of technical specification; data collection and processing; experimental data analysis.

M.K. Alenichev: concept advancement; data collection and processing; preparation of documents for reference material testing for the purpose of type approval.

Conflict of Interest The article was prepared on the basis of a report presented at the IV International Scientific Conference "Reference Materials in Measurement and Technology" (St. Petersburg, December 1–3, 2020). The article was admitted for publication after the abstract was revised, the article was formalized, and the review procedure was carried out.

The version in the Russian language is published in the journal "Measurement Standards. Reference Materials" 2021;17(1):59–69. https://doi.org/10.20915/2687-0886-2021-17-1-59-69.

References

1. Alenichev MK, Yushina AA, Drozhennikova EB, Filimonov IS, Levin AD, Baranova OA, Chekanov AV (2019) Fluorescent nanosensors based on colloidal quantum dots for the determination of reduced glutathione. Meas Tech 62(9):16–21. https://doi.org/10.1007/s11018-019-01695-x
2. Yaraki MT, Tan YN (2020) Recent Advances in metallic nanobiosensors development: colorimetric, dynamic light scattering and fluorescence detection. Sensors Int 100049. https://doi.org/10.1016/J.SINTL.2020.100049
3. Javidi M, Housaindokht MR, Verdian A, Razavizadeh BM (2018) Detection of chloramphenicol using a novel apta-sensing platform based on aptamer terminal-lock in milk samples. Anal Chim Acta 1039:116–123. https://doi.org/10.1016/j.aca.2018.07.041
4. Mi Y, Lei X, Han H, Liang J, Liu L (2018) A sensitive label-free FRET probe for glutathione based on CdSe/ZnS quantum dots and MnO2 nanosheets. Anal Methods 10(34):4170–4177. https://doi.org/10.1039/C8AY01532E
5. Ambrin G, Kausar H, Ahmad A (2020) Designing and construction of genetically encoded FRET-based nanosensor for qualitative analysis of digoxin. J Biotechnol 323:322–330. https://doi.org/10.1016/j.jbiotec.2020.09.008
6. Liu J, Bao C, Zhong X, Zhao C, Zhu L (2010) Highly selective detection of glutathione using a quantum-dot-based OFF–ON fluorescent probe. Chem Commun 46(17):2971–2973. https://doi.org/10.1039/b924299f
7. Yu L, Li L, Ding Y, Lu Y (2016) A fluorescent switch sensor for glutathione detection based on Mn-doped CdTe quantum dots-methyl viologen nanohybrids. J Fluoresc 26(2):651–660. https://doi.org/10.1007/s10895-015-1751-6
8. Wu D, Li G, Chen X, Qiu N, Shi X, Gl C et al (2017) Fluorometric determination and imaging of glutathione based on a thiol-triggered inner filter effect on the fluorescence of carbon dots. Microchim Acta 184(7):1923–1931. https://doi.org/10.1007/s00604-017-2187-2
9. Rezaei B, Shahshahanipour M, Ensafi AA (2017) In situ production of silver nanoparticles for high sensitive detection of ascorbic acid via inner filter effect. Mater Sci Eng C 71:663–668. https://doi.org/10.1016/j.msec.2016.10.046
10. Zheng T, Bott S, Huo Q (2016) Techniques for accurate sizing of gold nanoparticles using dynamic light scattering with particular application to chemical and biological sensing based on aggregate formation. ACS Appl Mater Interfaces 8(33):21585–21594. https://doi.org/10.1021/acsami.6b06903
11. Mustafaoglu N, Kiziltepe T, Bilgicer B (2017) Site-specific conjugation of an antibody on a gold nanoparticle surface for one-step diagnosis of prostate specific antigen with dynamic light scattering. Nanoscale 9(25):8684–8694. https://doi.org/10.1039/C7NR03096G
12. Zheng T, Finn C, Parrett CJ, Dhume K, Hwang JH, Sidhom D (2017) A rapid blood test to determine the active status and duration of acute viral infection. ACS Infect Dis 3(11):866–873. https://doi.org/10.1021/acsinfecdis.7b00137
13. Levin AD, Ringaci A, Alenichev MK, Drozhzhennikova EB, Shevchenko KG, Cherkasov VR, Nikitin MP, Nikitin PI (2020) Dynamic light scattering biosensing based on analyte-induced inhibition of nanoparticle aggregation. Anal Bioanal Chem 412:3423–3431. https://doi.org/10.1007/s00216-020-02605-9
14. Ulanova TS, Karnazhitskaya TD, Pshenichnikova YeO, Nakhieva EA (2013) The development of a method for chloramphenicol determination in meat products. Health Risk Anal (4):82–90. https://doi.org/10.21668/health.risk/2013.4.11
15. Alanazi AM, Mostafa GAE, Al-Badr AA (2015) Glutathione. In: Profiles of drug substances, excipients and related methodology (40):43–158
16. GOST ISO Guide 35-2015 (2016) Reference materials. General and statistical principles for certification. Standartinform Publ, Moscow, 61p (In Rus)
17. GOST ISO/IEC 17025-2019 (2019) General requirements for the competence of testing and calibration laboratories. Standartinform Publ, Moscow, 25p (In Rus)

18. State Pharmacopoeia of the Russian Federation. Edition XIV, Volume III, Medicines of synthetic and mineral origin. Federal Electronic Medical Library. http://femb.ru/femb/pharmacopea.php
19. GOST 5962-2013 (2014) Rectified ethyl alcohol from food raw materials. Technical conditions. Standartinform Publ, Moscow, 9p (In Rus)
20. RMG 93-2015 State system for ensuring the uniformity of measurements. Estimation of metrological characteristics of reference materials. https://beta.docs.cntd.ru/document/1200138923 (In Rus.)
21. GOST OIML R 76-1-2011 (2013) State system for ensuring the uniformity of measurements. Non-automatic weighing instruments. Part 1 Metrological and technical requirements tests. Standartinform Publ, Moscow, 140p (In Rus)
22. (2012) State system for ensuring the uniformity of measurements. Weights of classes E1, E2, F1, F2, M1, M1-2, M2, M2-3 and M3. Part 1 Metrological and technical requirements. Standartinform Publ, Moscow, 102p (In Rus)

Anna A. Yushina Engineer of the Laboratory of Analytical Spectroscopy and Nanoparticles Metrology, the All-Russian Scientific Research Institute for Optical and Physical Measurements.

Mikhail K. Alenichev Researcher of the Laboratory of Analytical Spectroscopy and Nanoparticles Metrology, the All-Russian Scientific Research Institute for Optical and Physical Measurements.

Correction to: Application of Lyophilization for Preparing Reference Materials of Food Composition

Maria Yu. Medvedevskikh, Natalia L. Vostrikova, Anna S. Sergeeva, and Valeriya V. Studenok

Correction to: Chapter "Application of Lyophilization for Preparing Reference Materials of Food Composition" in: S. V. Medvedevskikh et al. (eds.), *Reference Materials in Measurement and Technology*, https://doi.org/10.1007/978-3-031-06285-8_9

In the original version of the book, the following corrections have been incorporated:

In Reference 10 of chapter "Application of Lyophilization for Preparing Reference Materials of Food Composition", the author names have been corrected from "Saldanha H, Olsen SB, Ulberth F, Emons H, Zeleny R, Grimalt S" to "Grimalt S, Harbeck S, Shegunova P, Seghers J, Sejerøe-Olsen B, Emteborg H, Dabrio M". The book and the chapter have been updated with the changes.

The updated version of this chapter can be found at
https://doi.org/10.1007/978-3-031-06285-8_9

S. V. Medvedevskikh et al. (eds.), *Reference Materials in Measurement and Technology*,
https://doi.org/10.1007/978-3-031-06285-8_17